STP 1338

Superfund Risk Assessment in Soil Contamination Studies: Third Volume

Keith B. Hoddinott, editor

ASTM Stock #: STP1338

ASTM
100 Barr Harbor Drive
West Conshohocken, PA 19428-2959
Printed in the U.S.A.

Library of Congress Cataloging-in-Publication Data

Superfund risk assessment in soil contamination studies: third volume
/ Keith B. Hoddinott, editor.
   "ASTM Stock Number: STP1338"
   Papers presented at a symposium on Superfund risk assessment held
   in San Diego, California, on 11-16 January 1998.
   Includes bibliographical references and index.
   ISBN 0-8031-2484-8
      1. Hazardous wastes—Risk assessment—Congresses. 2. Soil
   pollution—Congresses. 3. Hazardous waste sites—United States—
   —Evaluation—Case studies—Congresses. I. Hoddinott, Keith B.,
   1956 -   . II. Series: ASTM special technical publication: 1338.
   TD1050.R57S87 1998
   363.739 ´62—dc21   98—42712
                                                                   CIP

Photocopy Rights

Peer Review Policy

Each paper published in this volume was evaluated by two peer reviewers and the editor. The authors
addressed all of the reviewers' comments to the satisfaction of both the technical editor(s) and the
ASTM Committee on Publications.
   To make technical information available as quickly as possible, the peer-reviewed papers in this
publication were prepared "camera-ready" as submitted by the authors.
   The quality of the papers in this publication reflects not only the obvious efforts of the authors and
the technical editor(s), but also the work of the peer reviewers. The ASTM Committee on Publications
acknowledges with appreciation their dedication and contribution of time and effort on behalf of ASTM.

Printed in Fredericksburg, VA
September 1998

Foreword

This publication, *Superfund Risk Assessment in Soil Contamination Studies: Third Volume,* contains papers presented at the symposium on Superfund Risk Assessment held in San Diego, California, on 11–16 January 1998. The symposium was sponsored by ASTM Committee D-18 on Soil and Rock. The symposium was chaired by Keith B. Hoddinott, U.S. Army Center for Health Promotion and Preventive Medicine. He also served as STP editor of this publication.

Contents

Overview

Since the beginning, there have been two schools of thought surrounding environmental risk assessment. One looked at the evaluation as an estimate of a true measure of risk an exposure would have on a population; the other viewed it as one of many tools that could be used to determine if exposure to a site and its chemicals posed an acute or long time risk to the health of the human and non-human receptors that would potentially use the site. Both schools have their advantages and drawbacks. The "true measure" school links the exposure to an effect that they believe is an estimate of reality, but they can get caught up in the Pandora's box of defending each value in the exposure model as how they describe reality at this site. The "one of many tools" school defends the exposure model as the standard measure for each site, but runs into controversy when evaluating one site against a set "safe" standard. Both schools suffer from the perception that they have to accept some level of adverse health effect. Both have merit in that if one has to make decisions concerning safety, a public health basis is an excellent place to start.

Public health has always been a science that attempts to keep ahead of the next problem. Refining techniques to detect problems earlier, make diagnoses more accurate, and get to the final solution quicker. Risk assessment is no different. From its inception, researchers have been looking closely at every facet of the risk assessment paradigm, ever tightening the uncertainty surrounding each parameter in the model. Over the years, numerous improvements and modifications have been made to the basic process. Keeping current with the changes is one of the more challenging parts of being a risk assessor.

The purpose of the symposium on risk assessment, which generated this Special Technical Publication (STP), was to collate the current modifications of the EPA's basic risk assessment methodology in a series of symposia and technical publications. We hope this type of symposium will serve both research and practical needs.

To produce this STP, two pro-active organizations combined their talents and resources. The American Society of Testing and Materials (ASTM), through its Committee D-18 on Soil and Rock, and the United States Army Center for Health Promotion and Preventive Medicine (formerly the United States Army Environmental Hygiene Agency) cosponsored the third of a series of symposia on this type of risk assessment.

The evaluation of these risks should follow the EPA's booklet entitled, "Risk Assessment Guidance for Superfund (RAGS)." This booklet outlines the general process of risk assessment which this STP has adopted to organize the paper topics. However, this STP does not pretend to be an instructional device for the basic EPA method. Although beginners can benefit greatly from the papers presented here, this collection finds its best use in the hands of the experienced risk assessor. The papers contained in the STP present modifications of the basic EPA methodology which have been acceptable to regulators at specific sites. This should not be construed to mean that these methods will be acceptable at all sites, in all situations, or to all regulators. Rather, it is a state-of-the-art laundry list of methods which may be helpful for complex issues at your site.

Papers in this STP were selected from the symposium submittals based upon pertinency, originality, and technical quality. All undergo peer review and most were extensively revised between presentation and publication.

In addition to the authors of the individual papers, any success of this publication reflects the contributions of many people. The Symposium Committee worked diligently in soliciting abstract submittals, in selecting promising presentations, and in chairing the sessions. The continued support of this symposium by the officers of ASTM Committee D-18 on Soil and Rock also was vital, as time from a more than full committee meeting schedule needed to be allocated for this endeavor.

Critical to maintaining the technical quality of this STP was the diligent work of the reviewers of the technical papers. At least three reviewers were obtained for each paper to help ensure that the work reported was accurate, reproducible, and meaningful.

Considerable staff support was also required for the completion of this effort. The help of the Symposium Committee, the D-18 officers, the paper reviewers, and the ASTM staff is most appreciated. We trust that the papers in this STP, which the contributors labored hard to develop, will aid the efforts of environmental professionals towards the reliable prediction and quantification of risk.

Keith Hoddinott

U.S. Army Center for Health Promotion
and Preventive Medicine
Aberdeen Proving Ground, MD;
Symposium Chairman and STP editor

Background Determination
and Statistics

Patrick D. Cook[1]

ESTIMATING BACKGROUND CONCENTRATIONS OF INORGANIC
ANALYTES FROM ON-SITE SOIL SAMPLE DATA

REFERENCE: Cook, P. D., **"Estimating Background Concentrations of Inorganic Analytes from On-Site Soil Sample Data,"** *Superfund Risk Assessment in Soil Contamination Studies: Third Volume, ASTM STP 1338*, K.B. Hoddinott, Ed., American Society for Testing and Materials, 1998.

ABSTRACT: At Superfund sites, regulatory agencies have commonly required that an extensive off-site study be performed to identify background concentrations of chemicals of concern in the sample medium for the purpose of establishing action levels. However, in an effort to save time and money, graphic statistics were used to estimate background concentrations of inorganic chemicals from 127 on-site soil samples collected at Andersen Air Force Base, a Superfund site on Guam. The statistical method used for this evaluation utilized probability plots, and was a modification of a technique used in mineral exploration to identify anomalous and background geochemical data. This paper provides an overview of the modified statistical method and its application at Andersen Air Force Base, and the results of the evaluation, including estimated background concentrations.

KEYWORDS: anomalous, arithmetic, background, logarithmic, population, probability plot, statistics, soil sample, threshold value

In remedial investigations (RIs), it is common to assume that soil samples collected from within the site are contaminated. This assumption is appropriate for many environmental sites. There are, however, numerous sites where contamination is suspected, but the location, nature, and extent of the contamination is either unknown or poorly defined. In these cases, there is a chance that some or possibly all on-site soil samples contain only background concentrations of inorganic chemicals.

To determine if on-site chemical data represent background or contamination, a background study is typically performed so that on-site concentrations can be compared to concentrations detected in areas considered to be unaffected by site operations. However, a background study is usually quite expensive and time-consuming. Therefore, an alternative method of distinguishing background concentrations from contamination

---

[1]Environmental Scientist, ICF Kaiser Engineers, Inc., 1600 West Carson St., Pittsburgh, PA 15219

was developed for Andersen Air Force Base (AFB). The method utilized probability plots to estimate background concentrations of inorganic analytes from on-site soil sample data which eliminated the need for a separate background study. Some benefits of using site data to estimate background rather than performing a separate study are listed below.

1)   Overall project costs can be reduced.

2)   Possible impacts from changes in sampling and analytical methodologies that could occur between the time background and site samples are collected are eliminated.

3)   Background definition can be enhanced because a) site data sets are generally larger than separate background data sets, and b) background data collected off-site may not be representative of site conditions.

Andersen AFB is a CERCLA (Comprehensive Environmental Response, Compensation and Liability Act) Superfund Site located on the Island of Guam, a U.S. Territory in the Pacific Ocean. The Base was placed on the National Priorities List (NPL) as a result of contamination detected several years ago in the underlying freshwater aquifer which supplies a considerable percentage of drinking water to military and civilian island residents. The CERCLA RI/FS (remedial investigation/feasibility study) included the characterization of the aquifer as well as numerous suspected disposal sites that were reportedly in operation from the 1940's through the 1970's. To facilitate the RI/FS, the disposal sites were grouped into operable units (OUs) based mainly on geography.

This paper involves an evaluation of data from one OU which consists of eight disposal sites (Table 1). Planning the investigation of these sites was difficult because very limited data were available on the sites' locations, waste types, and disposal methods. Most of the sites were heavily overgrown with thick jungle vegetation and were hard to distinguish from surrounding non-site areas. Preliminary investigations consisted of aerial photo interpretation, ground reconnaissance, geophysical surveys, and test excavations which were performed to provide the data needed to plan and perform a soil/waste sampling effort.

The samples listed below were collected during the RI/FS and were analyzed using U.S. Environmental Protection Agency (EPA) SW846 methods (EPA 1986) for the compounds and analytes on the Target Compound List and Target Analyte List.

- *Random Surface Soil Samples* - Collected at grid nodes to provide an unbiased estimate of chemical concentrations.
- *Biased Surface Soil Samples* - Collected at potential point sources to identify isolated areas of contamination.
- *Subsurface Soil Samples* - Collected immediately below waste to determine if the waste had impacted the underlying soils.
- *Subsurface Solid Waste Samples* - Collected in buried waste areas to determine if the wastes contained chemicals of concern.

TABLE 1--*Site descriptions*

Site	Site Type	Area (acres [hectares])	Waste Types
20	Area Landfill	1.84 [0.74]	Construction Debris
22A	Surface Dump	30.04 [12.16]	Miscellaneous Scattered Debris
22B	Drum Pile	0.03 [0.01]	Drums of Asphalt Tar
23	Trench Landfill	2.17 [0.88]	Sanitary Trash
24A	Surface Dump	9.87 [3.99]	Miscellaneous Scattered Debris
24B	Area Landfill	2.44 [0.99]	Sanitary Trash
37	Area Landfill	1.82 [0.74]	Automobile Parts
38	Spill Site	1.04 [0.42]	Dry Cleaning Fluids

Experimental Method

According to the work plan for the RI/FS, validated inorganic analytical data from site soil samples were to be screened against background concentrations which were to be established through a separate background study based on soil samples collected in areas considered to be unaffected by site waste disposal operations. After the work plan was finalized and field work initiated, Andersen AFB and the regulatory agencies agreed to take a different approach in evaluating the data. The approach required using on-site soil sample data to define inorganic background concentrations rather than performing a separate background study. To this end, a technique commonly used in mineral exploration for evaluating large sets of geochemical data (Sinclair 1976) was used to evaluate the Andersen AFB on-site soils data. The technique used cumulative probability plots, which are similar to cumulative histograms, to estimate threshold values between anomalous and background populations, and to identify apparent data distributions and outliers.

For every inorganic analyte, two probability plots were generated - one with an arithmetic scale, and the other with a logarithmic scale. The software used to create the probability plots allowed data from the appropriate soil samples to be readily graphed, whereas Sinclair (1976) described plotting class intervals of the data. Data from 60 random surface samples, 29 biased surface samples, and 38 subsurface samples were used to create the probability plots. Four solid waste samples and one sample contaminated with organic constituents were excluded from evaluation on probability plots to avoid any potential bias to threshold values. When an analyte was not detected in a sample, a value equal to one-half the reporting limit was included in the data set for that analyte, which was consistent with risk assessment practices used in the RI/FS.

The probability plots, some of which are described below, clearly indicated trends in analyte concentrations that were related to dominant soil types. The soils were all derived from the same parent material, and of the 127 soil samples included in the data set, 82 were generally classified as loams, 30 as sands, and 15 as clays. Although, the data set did not contain enough data points to create probability plots for each soil type, the relationship between soil type and analyte concentrations was still factored into the evaluation of the probability plots.

The probability (cumulative percentage) scale is arranged such that a cumulative normal or lognormal density distribution will plot as a straight line with relatively flat tails depending on whether the second axis of the probability paper is arithmetic or logarithmic (Fig. 1). Actual site data, however, did not always result in a single straight line when plotted on probability paper.

A curve with an apparent inflection point (point on the curve where a change of direction occurs) was commonly produced when the plotted data set contained multiple populations (Fig. 2). Such curves are referred to as polymodal probability curves.

Individual populations within some data sets were not easily distinguished by a sharp inflection point. This occurred when there was significant overlap between the individual populations. To more accurately determine where the populations separated, polymodal curves were partitioned (Sinclair 1976). Simply stated, partitioning is the extraction of the individual populations from the polymodal curve so that the parameters of each population can be evaluated separately. The partitioning of these populations was performed electronically. The first step was to visually identify the inflection point. Next, the data that existed at and below the inflection point were plotted separately. If this new plot still showed the presence of an inflection point (i.e., samples from the upper population were still evident), the data points above the inflection point were removed from the lower data set, and the remaining data were replotted. This process was continued until the individual populations were separated. Partitioned plots for certain analytes are presented below in the results discussion.

Multiple populations were evident in certain plots when the data set contained two or more background populations, anomalous populations, or a combination of the two. It is important to note however, that apparent multiple populations on several plots were related to several factors such as varying soil types (textures) and sample depths, spatial distribution, and the presence of qualified data.

To help determine if the above factors affected the shape of a curve, additional sample information was used during the review of the probability plots to help explain curve behavior and to help avoid false conclusions. Side graphs indicating site location,

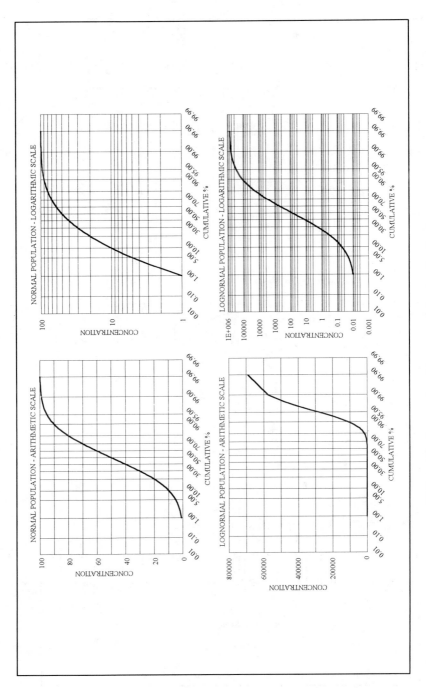

FIG. 1--*Ideal normal and lognormal density distributions.*

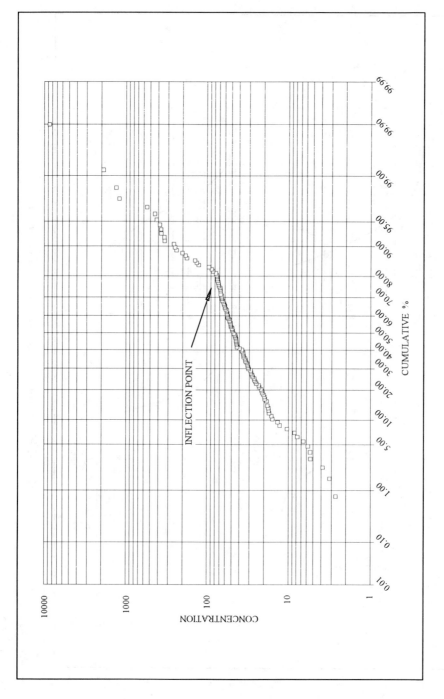

FIG. 2–Multiple population probability plot.

sample type, soil type, and qualification status were placed adjacent to the probability plot (Fig. 3). The left side graphs were used to evaluate whether or not an apparent inflection point was related to effects from sample types (surface vs. subsurface, and biased vs. random), soil types (e.g., clay vs. sand), or the presence of qualified data (i.e., nondetects and estimated concentrations). The right side graphs were used to spatially analyze the data for the presence of outliers which could also affect the shape of the curve. Inter-site outliers were identified between sites when the concentrations of an analyte in one site's data set were significantly different from the other sites' data sets. Intra-site outliers were identified when one or more data points within an individual site data set were significantly different from the other data points in the same set.

In addition to the side graphs, field records, analytical data, and general site knowledge were instrumental in evaluating the probability plots. The evaluation of several probability plots is presented later in the results discussion.

A systematic process was used for evaluating probability plots to estimate background threshold values. When a plot showed the presence of only one population, the data were assumed to represent background unless site knowledge indicated that the single population represented contamination. The threshold value was arbitrarily selected at the concentration that corresponded to the 95th percentile of the data. This was a conservative measure aimed at locating anomalous values that may have overlapped into the background population. The upper 5 percent of the data were further evaluated (i.e., examining field and laboratory documentation) to help confirm that these samples in the higher concentration range were not anomalous. Note that any sample in the lower 95 percent of the data that appeared to be an outlier for some reason was also further evaluated in the same manner as samples in the upper 5 percent.

When a plot indicated the presence of multiple populations, the side graphs adjacent to the plot were examined to determine if the change in the slope of the curve was caused by the factors discussed above. If the side graphs indicated that the appearance of multiple populations was due to changes in soil type or sample type, and/or a grouping of qualified data, then it was assumed that the entire data set represented background. In these cases, the selection of the background threshold value was performed as described above for a single (background) population.

Conversely, if the side graphs and site knowledge did not explain the presence of multiple populations, then a potentially contaminated population was assumed to be present. If the contaminated population did not extend down into the lower 95th percentile of the data, the selection of the background threshold value and the further evaluation of the contaminated data set were performed as described above for a single (background) population.

If a probability plot contained an identified contaminated population that extended

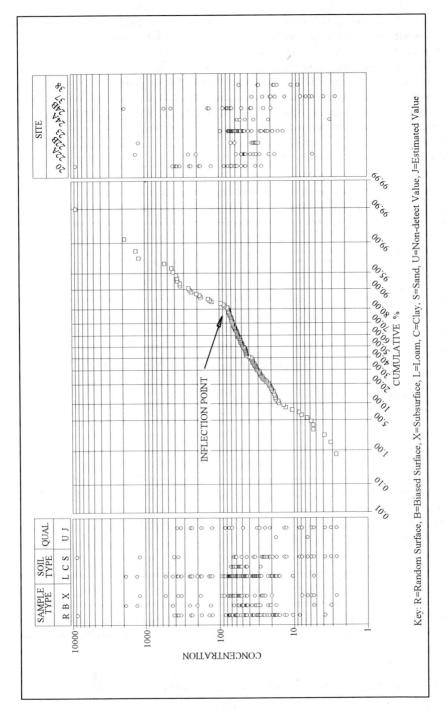

FIG. 3--Multiple population probability plot with side graphs.

Key: R=Random Surface, B=Biased Surface, X=Subsurface, L=Loam, C=Clay, S=Sand, U=Non-detect Value, J=Estimated Value

below the 95th percentile, the plot was partitioned to separate the background and potentially contaminated populations. After partitioning, the selection of the background threshold value and the further evaluation of the data set were performed on the partitioned background population as described above for a single population. The samples in the contaminated population were also subjected to further evaluation in the same manner.

Results

The probability plot evaluation produced an estimated background threshold value for each analyte (Table 2). The probability plots showed that six of the 18 inorganic analyte data sets had both background and anomalous populations. The remaining analyte data sets only had background populations. Most background populations exhibited a normal distribution, whereas anomalous populations were generally lognormal. However, there were exceptions (Table 2).

The selection of background threshold values was straightforward for some analytes, and more complex for others because different population scenarios were encountered during the review of the plots. Some analyte data sets contained only background populations, while others some contained background and contaminated populations. Some contaminated populations extended below the 95th percentile and others did not. The process for evaluating the plots and estimating background threshold values for the different population scenarios was discussed previously. The selection of background threshold values for aluminum (Al), cadmium (Cd), and zinc (Zn) are discussed below and represent the various population scenarios encountered during the review of the probability plots.

Aluminum

The Al concentrations were plotted on arithmetic and logarithmic probability paper (Fig. 4 and Fig. 5, respectively). These plots showed the presence of only one population which exhibited a normal density distribution. The arithmetic probability plot (Fig. 4) showed a slight change in the slope of the curve at about 75,000 mg/kg. However, the change in slope was not considered an inflection point between background and anomalous populations because the left side graph indicated the change was likely related to soil texture. Most of the samples with a high sand content had Al concentrations below 75,000 mg/kg, whereas samples with a high clay content generally contained much higher Al levels. This relationship between soil texture and analyte concentration was observed for other metals in this study as well. Typically, trace elements were found at higher levels in clayey soils, while the concentration of major elements were highest in sandy soils. This relationship is consistent with observations reported by Kabata-Pendias (1984). However, there were exceptions. For example, Al which is a major element, was detected in clayey soils at concentrations higher than the levels detected in sandy soils.

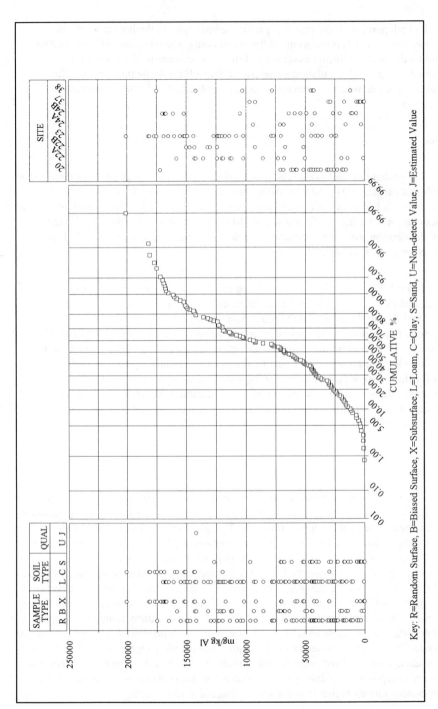

FIG. 4--*Arithmetic probability plot of aluminum data.*

Key: R=Random Surface, B=Biased Surface, X=Subsurface, L=Loam, C=Clay, S=Sand, U=Non-detect Value, J=Estimated Value

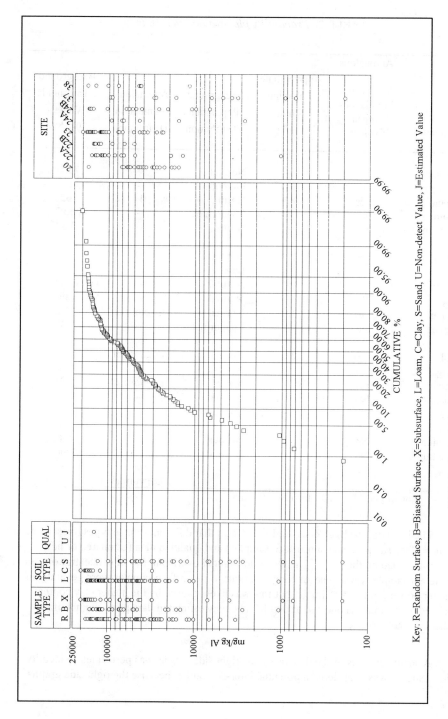

FIG. 5--*Logarithmic probability plot of aluminum data.*

Key: R=Random Surface, B=Biased Surface, X=Subsurface, L=Loam, C=Clay, S=Sand, U=Non-detect Value, J=Estimated Value

TABLE 2--*Probability plot evaluation results*

Parameter	Anomalous and Background Populations Appear Evident?	Approximate Inflection Point (mg/kg)	Background Population Distribution	Anomalous Population Distribution	Background Threshold Value (mg/kg)
Aluminum	No	--	Normal	--	173,500
Antimony	No	--	Lognormal	--	63
Arsenic	Yes	82	Normal	ID	62
Barium	No	--	Lognormal	--	335
Beryllium	No	--	Normal	--	3.34
Cadmium	Yes	7.1	Normal & Lognormal	ID	6.5
Chromium	No	--	Normal	--	1,080
Cobalt	No	--	Normal	--	29
Copper	Yes	80	Normal	Lognormal	72.2
Cyanide	No	--	Lognormal	--	1.47
Lead	Yes	220	Normal	Lognormal	166
Manganese	No	--	Normal	--	3,150
Mercury	Yes	0.345	Normal	Lognormal	0.28
Nickel	No	--	Normal	--	243
Silver	No	--	Lognormal	--	14.9
Thallium	No	--	Normal	--	1.42
Vanadium	No	--	Normal	--	206
Zinc	Yes	140	Normal	Normal & Lognormal	111

Key: -- = Not Applicable,   ID= Insufficient Data

The left side graphs also showed the effects that different sample types (biased and random surface samples, and subsurface samples) and qualified data (estimated or non-detect values) had on the curve. There was substantial overlap in the Al concentrations for the different sample types. The Al concentrations in subsurface samples were generally in the upper portion of the data range, but this was likely caused by the higher clay content associated with samples collected at depth. The only estimated data point did not affect the shape of the curve.

A spatial analysis of the data using the right side graphs was performed to identify outliers. Site 23 was considered a potential inter-site outlier because the right side graphs showed that most of the samples containing more than 150,000 mg/kg Al were taken at this site. However, because the field records indicated that these soil samples consisted of clay, the relatively high Al levels at Site 23 were not considered anomalous. The right side graphs also showed a potential intra-site outlier within the Site 20 data set. The sample analytical data (not included in this paper) indicated that most of the Site 20 samples contained less than 71,000 mg/kg Al, but one sample contained 172,000 mg/kg Al. The sample containing 172,000 mg/kg Al was not considered anomalous because the field records indicated that this sample had a high clay content relative to the other Site 20 samples.

Because the Al probability plots showed only one population, and the side graphs did not provide evidence to suggest the presence of contamination, the entire data set was considered to represent background. However, rather than immediately concluding that all Al data represented background, the background threshold value was selected at the concentration corresponding to the 95th percentile of the data (173,500 mg/kg Al). As a conservative measure, the upper 5 percent of the data were further evaluated to help confirm that these values did not represent contamination. The further evaluation included the examination of field and laboratory documentation, and a correlation analysis with other analytes. The evaluation did not indicate that the upper 5 percent were anomalous. Similar steps were taken for other metals that did not exhibit an upper population (Table 2).

Cadmium

The arithmetic and logarithmic probability plots of Cd data (Fig. 6 and Fig. 7, respectively) indicated the presence of two Cd populations. Partitioning of the plots revealed an inflection point at about 7.1 mg/kg and relatively few data points in the upper population. The left side graphs showed substantial overlap of Cd concentrations relative to the different soil types and sample types. Therefore, it was considered unlikely that soil texture or sample type differences produced the inflection point. Although 21 non-detect values existed in the lower end of the concentration range (less than 1.75 mg/kg), these data did not appear to significantly alter the curve. Nevertheless, these nondetect data appeared as a separate group because they were assigned a value of one-half the reporting limit.

The right side graphs showed substantial overlap in Cd concentrations between the sites, thus there were no inter-site outliers. There were, however, several intra-site outliers above the inflection point as shown on the right side graphs. For example, most of the Site 22B samples contained less than 7 mg/kg Cd, yet one sample contained 183 mg/kg Cd. The high concentration did not appear related to soil type. Field records indicated that the sample was taken from under metal debris which was the likely source of the 183 mg/kg Cd concentration. The review of field records was vital to the interpretation of the Cd probability plots as well as plots of other analytes.

16 SUPERFUND RISK ASSESSMENT: THIRD VOLUME

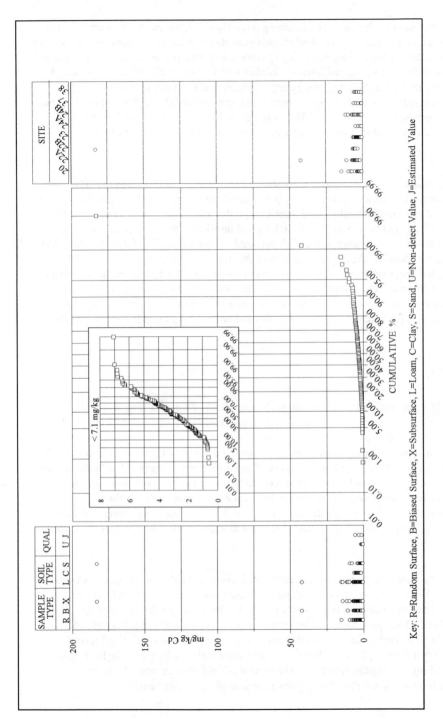

FIG. 6--*Arithmetic probability plot of cadmium data.*

Key: R=Random Surface, B=Biased Surface, X=Subsurface, L=Loam, C=Clay, S=Sand, U=Non-detect Value, J=Estimated Value

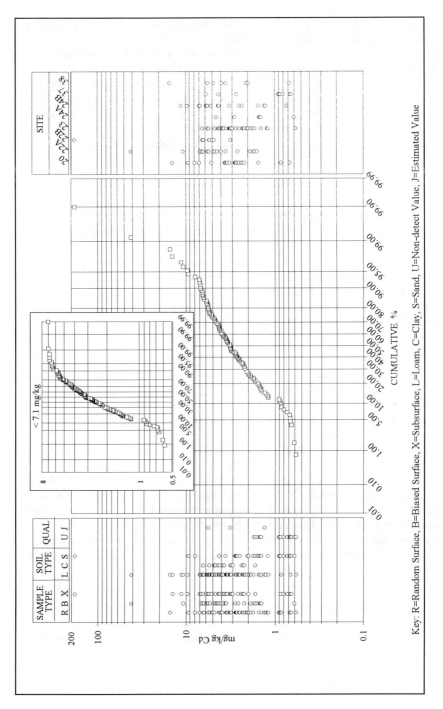

FIG. 7--Logarithmic probability plot of cadmium data.

Key: R=Random Surface, B=Biased Surface, X=Subsurface, L=Loam, C=Clay, S=Sand, U=Non-detect Value, J=Estimated Value

The upper Cd population was considered anomalous because the side graphs and site knowledge provided no evidence to the contrary. Therefore, to establish the background threshold value for Cd, the populations were partitioned, as described by Sinclair (1976), to evaluate the background data independently. The partitioned background population (Fig. 6 and Fig. 7 inset plots) exhibited normal and lognormal density distributions. The upper population did not contain enough data points to plot separately; thus, the distribution of the contaminated population could not be determined from probability plots. The background threshold value was set at the concentration corresponding to the 95th percentile of the background population (6.5 mg/kg Cd).

Zinc

The Zn arithmetic and logarithmic probability plots (Fig. 8 and Fig. 9, respectively) indicated the presence of multiple populations with an apparent inflection point at about 140 mg/kg. The data curve was partitioned to separate the multiple populations. Initially, the probability plot of the entire Zn data set indicated the potential for two populations above 140 mg/kg and one below. However, partitioning of the plots revealed only one population above 140 mg/kg. The inset plots (Fig. 8 and Fig. 9) indicated that both populations were normally distributed, and that the upper population was also possibly lognormally distributed.

Unlike Cd, the upper Zn population contained numerous data points. The different sample types and the presence of 15 estimated values did not appear to significantly affect the shape of the curve. However, the side graphs (Fig. 8 and Fig. 9) and field records provided compelling evidence that the upper population was anomalous. The right side graphs showed that Site 20 was an inter-site outlier because all but one of the samples taken at that site had Zn concentrations above the 140 mg/kg inflection point. At other sites, most of the Zn levels were below the inflection point. Field records revealed that Site 20 contained metal construction debris throughout the surface and subsurface soil strata, which may have been galvanized and was the likely source of the elevated Zn concentrations. Further, the left side graphs and partitioned plots indicated that clayey samples were grouped near the top of the concentration range in the lower or background population (0 to 140 mg/kg). Given that the clayey samples in this investigation were found to have relatively high natural levels of Zn, the low percentage of clayey samples in the upper population (above 140 mg/kg) suggested the upper population did not represent background.

Intra-site outliers were also identified on the right side graphs and explained by field records. For example, several samples taken at Site 22A and Site 24B had relatively high levels of Zn. Field records showed that these samples were collected near surficial metal debris.

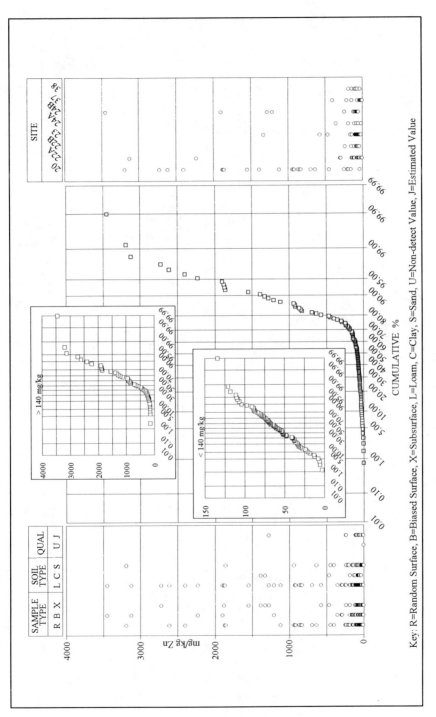

FIG. 8--*Arithmetic probability plot of zinc data.*

Key: R=Random Surface, B=Biased Surface, X=Subsurface, L=Loam, C=Clay, S=Sand, U=Non-detect Value, J=Estimated Value

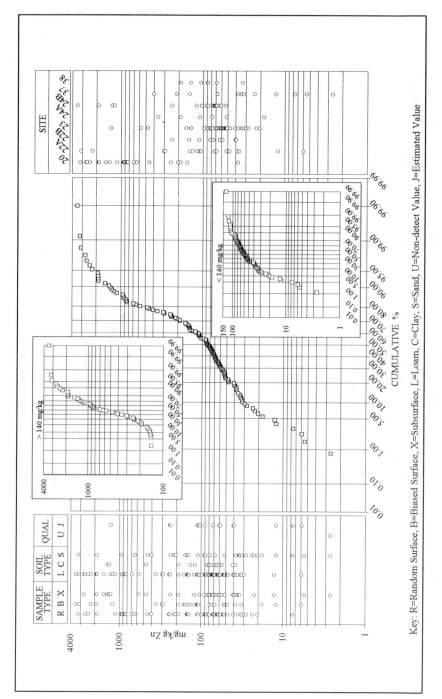

FIG. 9--*Logarithmic probability plot of zinc data.*

Key: R=Random Surface, B=Biased Surface, X=Subsurface, L=Loam, C=Clay, S=Sand, U=Non-detect Value, J=Estimated Value

The background threshold value for Zn (111 mg/kg) was set at the concentration corresponding to the 95th percentile of the lower or background population.

Summary and Conclusions

The remedial investigation of several sites at Andersen AFB was performed without spending time and resources on a separate background study because probability plots, based on Sinclair (1976), were used to estimate background concentrations of inorganic analytes from on-site soil sample data. At Andersen AFB, the nature and extent of contamination was poorly defined and it was considered likely that many of the samples taken to check for the presence of contamination would only contain background concentrations of inorganic analytes. However, even if most of the data represented contamination, this method would still be successful in estimating background threshold values.

Estimating background threshold values from probability plots was a subjective process. To help avoid false conclusions, careful consideration was given to actual site conditions, sample descriptions, spatial distribution, and the degree to which different soil types, sample types, or qualified data affected the appearance of the plot. Graphs of sample information placed adjacent to the probability plots, as well as field records, factored into the background threshold selection process, and the identification of outliers. Without the side graphs, field records and analytical data, it would have been very difficult to evaluate the data and estimate background threshold values.

For many analytes, the selection of a background threshold value from a probability plot was straightforward because the data set only contained a background population. The evaluation of other analytes was more complex because multiple populations were present in the same data set. In some cases, an apparent inflection point was observed on a plot but was considered related to soil type rather than the presence of anomalous data.

One-third of the inorganic analyte data sets had both background and anomalous populations. The remaining analyte data sets had only background populations. Multiple populations were partitioned to allow independent evaluation; however, some anomalous populations did not contain enough data points to plot separately. The probability plots indicated that background populations were usually normally distributed, and anomalous populations were generally lognormally distributed; however, there were exceptions.

Although this method was only used for inorganics at Andersen AFB, it could be used to estimate background threshold levels for other chemicals such as polynuclear aromatic hydrocarbons (PAHs) that may be present in soils but are not site-related.

Acknowledgements

The author would like to acknowledge and express gratitude to John Miller of Mitretek Systems for suggesting the use of probability plots, and Kenny Smith of ICF Kaiser Engineers, Inc. for providing computer software technical support.

References

Kabata-Pendias, A., and Pendias, H. 1984, *Trace Elements in Soils and Plants*. CRC Press, Inc., Florida.

Sinclair, Alastair J., 1976, "Applications of Probability Graphs in Mineral Exploration," *The Association of Exploration Geochemists,* Special Vol. No.4, pp. 1-95, Richmond Printers LTD, Richmond, B.C. Canada.

United States Environmental Protection Agency, 1986, *SW846 Test Methods for Evaluating Solid Waste*, Third Edition, Government Printing Office, Washington, DC.

Roy O. Ball[1] and Melinda W. Hahn[1]

THE STATISTICS OF SMALL DATA SETS

REFERENCE: Ball, R. O., and Hahn, M. W., "**The Statistics of Small Data Sets**", *Superfund Risk Assessment in Soil Contamination Studies: Third Volume, ASTM STP 1338*, K.B. Hoddinott, Ed., American Society for Testing and Materials, 1998.

ABSTRACT: A primary concern in Risk-Based Corrective Action (RBCA) is the decision criteria for evaluating attainment of cleanup objectives. Carcinogenic compounds (with long modeled exposure duration) typically drive remediation at most sites. The statistic for comparison with the risk-based cleanup objective should therefore be an upper bound estimate (the "Reasonable Maximum Exposure" [RME]) for the true spatially averaged concentration. Contaminant concentrations in soil are inherently lognormally distributed. Small simulated "sample sets" selected at random from a lognormal distribution were used to test the accuracy and stability of the statistical method recommended in Superfund guidance and that recommended in RCRA SW-846 (using the log transformation). The results show that the geomean is the more accurate and stable estimator of true average exposure concentration and that the 90% Upper Confidence Limit of the geomean is a conservative statistic for comparison with risk-based standards.

KEYWORDS: environmental data, statistical analysis, Monte Carlo, mean, variance, lognormal distribution, cleanup objective

The Land Pollution Control practitioner faces five significant issues in the RBCA process:

1: How are the site Land Pollution issues defined and delineated?
2: How many data are needed to reliably characterize the three-dimensional Tier I Risk-Based Screening Level (RBSL) boundaries?
3: Which statistical distributions and statistics best describe Land Pollution site data?

---

[1] ENVIRON International

4: Which statistical distributions and statistics best describe fate, transport & toxicological variables?
5: What decision criteria and statistics should be used to compare site data to RBCA Site Specific Target Levels (SSTLs)?

The overall objective of this paper is to propose solutions to the third issue. The specific objectives are to discuss the special case of statistical properties of "small data sets", to compare the statistical analysis methodologies of RCRA SW-846 & Superfund RAGs, to propose decision criteria to compare site data to SSTLs, and to illustrate the proposed methodology using hypothetical site data.

To accomplish these objectives, the paper is divided into four sections. Section I provides a literature review on the statistics of small data sets - including discussion of current US EPA guidance for RCRA (SW-846, 1986) and Superfund (RAGs, 1992). Section II describes the statistical properties of environmental releases, and demonstrates that the population geomean is a reliable estimator of the true spatially averaged concentration (which is the best descriptor of environmental exposure). Section III examines the stability and variability of small data set statistics as a function of sample size, and Section IV provides conclusions and recommendation for statistical decision criteria for small data sets.

Section I – Literature Review

Environmental risk is calculated as the product of toxicology factors and dosage. Dosage is generally expressed as the mass of contaminant ingested, inhaled or otherwise absorbed, divided by the receptor mass per unit time. Therefore, the concentration term is directly proportional to the dosage or exposure, and to the corresponding risk. There are significant economic consequences of excessive conservatism in calculating the site concentration term for risk assessment, without any corresponding benefit to public health. Therefore, calculating a reliable estimate for the true average exposure concentration is important.

According to the May, 1992 Supplemental Guidance to RAGs, "… the concentration term C in the intake equation is an estimate of the … average concentration for a contaminant based on a set of site sampling results." An estimate of average concentration is used because it "…is most representative of the concentration that would be contacted at a site over time"-"…based on lifetime average exposures…"and "…the spatially averaged soil concentration can be used to estimate the true average concentration contacted over time." Therefore, the problem of reliably determining the environmental risk due to site contamination is directly related to, *inter alia*, estimating the spatially averaged soil concentration, generally using small data sets. 'Small data sets' are not precisely defined in the literature, but generally are taken to mean less than 30 samples per exposure (or source) area.[2]

Current textbooks on statistics for environmental problems do not directly address the issue of small data set variability. As an example, the textbook generally referenced

---

[2] For example, the RAGs Supplemental Guidance of May, 1992, states that, "…fewer than 10 samples … provide poor estimates of the mean concentration … while data sets with 20 to 30 samples provide … consistent estimates of the mean"

in the SuperFund Manuals and Guidance (Gilbert 1987) in the Chapter on "Characterizing Lognormal Populations", provides a mathematically correct account of the properties of various statistics without discussion of this issue. Berthouex and Brown (1994), note that there are good reasons to transform environmental data for statistical analysis, but cautions that if the natural log transformation has been used, "…the simple back-transformation of antilog (data) does not give an unbiased estimate of… (the original distribution)" without exploring the adequacy of the back-transformed statistic, or any other statistic, to reliably predict the true spatially-averaged concentration. Ott (1995) provides a very helpful discussion of the nature of environmental release to create lognormal or near-lognormal distributions[3], and that simple observation of graphical displays of small data set distributions is a reasonable procedure for assigning a statistical distribution to the data set. However, the discussion of small data set statistics is primarily mathematical.

The concern of predicting the spatially averaged concentration with realistic precision (given the economic consequences otherwise) is more evident in the literature. Gleit (1985) showed that the methods used to 'fill-in' censored data made "…orders of magnitude" differences, and recommended the use of expected values based on regression. Gilliom and Helsel (1986) and Helsel and Gilliom (1986) also provided experimental evidence of the wild inaccuracies of some small data set statistics (including those routinely used for regulatory purposes). They recommended lognormal regression of the uncensored data to estimate censored values as the most robust estimator of the population mean (as well as of other statistics). Shumway *et al.* (1989) recognized that "…a method should be sought for transforming the estimators of the mean in the transformed scale back to an estimator … in the original scale." They also concluded that "When environmental data cannot be modeled …(by)… the normal distribution, the (arithmetic) sample mean is not … the best estimator for the population mean."[4] Haas (1990) in describing a method to 'fill-in' censored data that was robust to non-normality, provided further experimental results demonstrating the extreme bias and error of many small data set statistics. As a result, it is difficult to understand the apparent lack of understanding presented in two of the principal references used in SuperFund for contaminated site assessment (Guidance for conducting RI/FS under CERCLA and RAGs). These standard references either provide minimal treatment of the subject (Guidance) or recommend the use of an excessively variable and conservative statistic (RAGs). In the RCRA program, Volume II of SW-846 does provide appropriate treatment of this topic. However, even here, the recommendations are presented without extensive discussion or justification.

In summary, there is at least implicit agreement in the literature that the variability associated with the statistical properties of small data sets is best examined by experiment. The results of such experiments demonstrate that small data set statistics can be significantly biased and inaccurate.

---

[3] Also described in J. Air Waste Manage. Assoc.,**40**, 1378-1383 (1990)
[4] This directly contradicts the RAGs Supplemental Guidance of May, 1992, which incorrectly states, "…the arithmetic mean …(is)… the appropriate measure for estimating exposure…", even though EPA correctly recognizes the need to estimate the true spatially-averaged soil concentration.

Section II – Statistical Properties of Environmental Releases

The problems in land pollution control are usually due to an accidental spill or release. Such generating events inevitably create a lognormal or near-lognormal distribution of contamination (Ott 1990), as the contaminant pattern in space and time is caused by successive dilution. Even if there is doubt about the underlying distribution of the contamination, Berthouex and Brown (1994) conclude that, "… making the lognormal transformation is usually beneficial, or at worst harmless." [5] This also avoids the problem discussed by Shumway *et al.* (1989), "…for most environmental data, the assumption that the underlying distribution is normal will not be appropriate, so the usual sample mean" (*i.e.*, the arithmetic mean) "will not be a good estimator of the population mean." This process of dilution and dispersion of a concentrated source inevitably creates an auto-correlated, or Markovian, distribution. Therefore, while the statistical evaluation of small data sets can provide a reliable estimate of the mean and reasonable upper bound of the contaminant exposure, the most reliable treatment of land pollution data is to apply geo-statistical analysis, or kriging. Kriging can define the maximum and minimum characteristic distances of autocorrelation as well as illuminating the effect of media anisotropy on contaminant distribution. Because kriging, particularly in three dimensions, is more resource intensive than the simple statistical analyses described herein, it is most reasonably applied where statistical analysis indicates the effort is warranted.

The following hypothetical release results in lognormally distributed soil contamination. The contents of a trichloroethene (TCE) drum were assumed to have been released, such that the source concentration was equal to 100 mg/kg. The distribution of contamination from the source is arbitrarily defined as:

$$C_{rz} \equiv C_o \, e^{-(z/\vartheta_z + r/\vartheta_r)} \tag{1}$$

Where z is the depth from the ground surface, r is the radial distance from the release centroid and the release attenuation parameters (θ_r and θ_z) define the characteristic distance of attenuation along each axis.

This exponential decay equation provides a realistic distribution of contamination. In actuality, such distributions do have a random component due to variations of soil properties and the rate and conditions of contaminant release. The incorporation of a random component would not change the results of the experiment, but would add some complexity. Therefore, the experiment herein assumes isotropic dispersion. By mathematically defining the release, it is possible to accurately benchmark the true contaminant distribution, to calculate the true spatially averaged contaminant concentration, and to generate small data set sampling results (and statistics) for any defined sampling methodology.

---

[5] Many other authors concur with Berthouex and Brown *(e.g.*, Newman *et al.* 1989, etc.). Stoline (1990), however, cautions against, "…the automatic use of the lognormal model…" and recommends, "…checking the adequacy of the lognormal model prior to use."

For this exponentially attenuated release, the incremental volume and mass of any soil element can be calculated by:

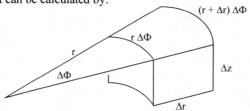

$$\Delta V = (r\Delta\Phi + \Delta r\Delta\Phi / 2)\Delta r\Delta z = (r\Delta r + \Delta r^2 / 2)\Delta\Phi\Delta z \tag{2}$$

and

$$\Delta M = \rho_s C_o C_{ra}\Delta V \tag{3}$$

respectively.

Therefore, the total volume and mass are given by:

$$V = \sum_{\varphi} \sum_{r} \sum_{z} \Delta V \qquad M = \sum_{\varphi} \sum_{r} \sum_{z} \Delta M \tag{4}$$

As discussed in Section I, the RAGs Supplemental Guidance (1992) has established that the spatially averaged concentration is the appropriate statistic for comparison with risk-based standards because 1) carcinogenic and chronic non-carcinogenic toxicity criteria are based on lifetime average exposures, and 2) the average concentration is most representative of the concentration that would be contacted over time (the average exposure concentration). The true spatially averaged concentration of a contaminant in the release will be:

$$\bar{C} = M / V \tag{5}$$

Simulated contaminant concentrations were calculated for one-foot intervals of [r, z, Φ] up to [50,20,2π] using Equation 1 with θ_r = 8 feet and θ_z = 5 feet. The results were truncated at values of C/C_o equal to 0.1%). Using Equations 2 through 5, the true spatially averaged concentration was calculated to be 1.2 mg/kg for a release source strength of 100 mg/kg. The data (for the normalized concentrations) are well described by a two-parameter lognormal distribution with mean and standard deviation of {-4.45, 2.08}. Therefore, for C_0 = 100 mg/kg, the true spatially averaged concentration is also equal to $100 \times e^{-4.45}$ or 1.2 mg/kg, the population geomean. This indicates that the population geomean (or log-transformed mean) is an unbiased and accurate measure of the true spatially averaged concentration.

The following sampling strategy was applied to investigate the hypothetical release of TCE. Based on organic vapor meter screening, it was determined that the release occupies a volume of approximately 37 ft in radius with a penetration depth of 20 feet. Eight (8) samples were collected within this cylinder, at randomly selected coordinates for [r, z].

Each group of eight samples constitutes a "small data set" representing lognormally distributed soil contamination. The properties of such small data sets are best displayed by Monte Carlo analysis, where many hypothetical sample sets can be generated and analyzed. A sample "trial" is defined as a set of 8 data, collected randomly from the true lognormal contaminant probability distribution function, or pdf, (*i.e.*, the contaminant distribution created by Equation 1 with θ_r = 8 ft and θ_z = 5ft) with transformed mean and standard deviation of -4.45 and 2.08, respectively. Random, representative samples were created by use of the Excel® spreadsheet function "***LOGINV[probability, mean, standard deviation]***"[6] For the pdf described above, the function is: "***LOGINV[RAND(),-4.45, 2.08]***". By replicating this function, any desired number of random, representative samples can be created.

For each sampling trial, the following statistics were calculated:

- arithmetic (or untransformed) mean
$$\mu = \sum_{i=1}^{n} x_i / n$$

- arithmetic standard deviation
$$\sigma = \sqrt{\frac{\sum_{i=1}^{n}(x_i - \mu)^2}{n-1}}$$

- arithmetic coefficient of variation $\eta = \mu / \sigma$

- 95% UCL[7] of the arithmetic mean[8]:   "***CONFIDENCE[1-.95, σ, n]***"

- arithmetic maximum x_{max}

- geomean
$$\mu_g = \exp\left(\sum_{i=1}^{n} y_i / n\right);\ldots\ldots y_i = \ln(x_i)$$

- log-transformed standard deviation[9]
$$\sigma_g = \exp\left(\sqrt{\frac{\sum_{i=1}^{n}(y_i - \mu)^2}{n-1}}\right)$$

- 90% UCL of the geomean $\exp($"***(CONFIDENCE[1-.9, ln(σ_g), n]***"$)$

A total of 2,500 trials were run. The true spatially averaged concentration of the lognormal distribution (equal to the population geomean) and the 90% UCL of the true spatially averaged concentration (equal to the 90% UCL of the population geomean) were

---

[6] Hereafter spreadsheet functions will be denoted as, "***FUNCTION[parameter 1, 2,..., n]***"
[7] UCL= Upper Confidence Limit
[8] Assumes underlying distribution is normal
[9] here μ represents the mean of the y_i

used as "benchmarks" for the sample statistics described above. The following ratios were also calculated:

- Sample mean (average)/population geomean concentration[10]
- Sample RME_{RAGs} /90% UCL of the population geomean
- Sample geomean/population geomean concentration
- Sample RME_{SW-846} /90% UCL of the population geomean

where the sample RME_{RAGs} is the lesser of the sample maximum and the sample 95% UCL of the arithmetic mean as described in Superfund guidance, and the RME_{SW-846} is equal to the 90% UCL of the sample geomean as described in SW-846, Volume II.

Table 1 shows the average of the statistics for the Monte Carlo trials, and Figures 1A and 1B show the histogram for the four ratios defined above. Clearly, the sample geomean more accurately describes the true average concentration as evidenced by the sharp peak in frequency between values of the ratio between 0.5 and 1.5, and the sharp decrease in frequency above 4. The frequency of the sample mean, on the other hand, peaks between the values of 3 and 4, and has a pronounced tail that suggests that the arithmetic mean may overestimate the true average by up to a factor of approximately 20. Similarly, the sample RME_{SW-846} is the more accurate and stable estimator of the upper bound of the true average concentration compared to the sample RME_{RAGs}. The sample RME_{RAGs} can overestimate the 90% UCL of the population distribution geomean by up to a factor of approximately 50.

TABLE 1—*Average of sample statistics.*

	Mean	StDev	η	95% UCL	Max	MRE
Arithmetic	5.2	8.0	1.5	10.7	22.6	10.7

	Mean	StDev	90% UCL
Log-Transformed	1.6	6.1	4.4

---

[10] The population geomean equals the spatially averaged concentration. Hereafter, the terms are synonymous and will be used interchangeably

Figure 1-A --
*Variation of Sample Statistics
for Small Data Sets [full scale]*

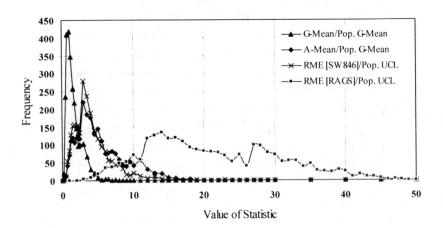

Value of Statistic

Figure 1-B --
*Variation of Sample Statistics
for Small Data Sets [reduced scale]*

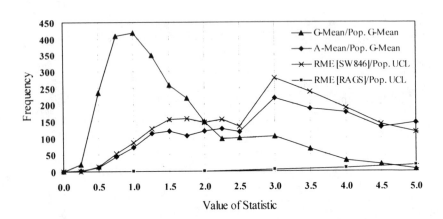

Value of Statistic

Section III – Statistical Variation due to Sample Size

Using the lognormal probability density function (pdf) for the hypothetical release of TCE described in Section II, random, representative samples were again created by use of the Excel® spreadsheet function "*LOGINV[RAND(),-4.45, 2.08]*". The "*LOGINV[RAND(),-4.45, 2.08]*" function was used to create small data sets of three (3) to thirty (30) samples. For each sample set, the same statistics listed in Section II were calculated. To better evaluate the stability of the calculated statistics from small data sets, larger sample sets with n = 50, 75, 100, 250, 500 and 1000 were also created.

Figure 2 compares the results from a "trial"[11] for variation in arithmetic mean[12] and geomean for sample sizes from 3 to 30. (The trial results have been divided by the true average concentration of 1.2 mg/kg for the lognormal pdf). Note that the arithmetic mean exhibits significantly greater variability, particularly for the smaller sample sizes, and appears to fluctuate around a ratio of approximately 5. The geomean, except for the smaller sample sizes, fluctuates around ratio of approximately 1.5, without apparent bias. This indicates that the log-transformed, or geomean, is a conservative and stable estimator of the true average concentration,[13] even for small data sets.

Figure 2 --

Variation of Sample Means with Sample Size
for Small Data Sets

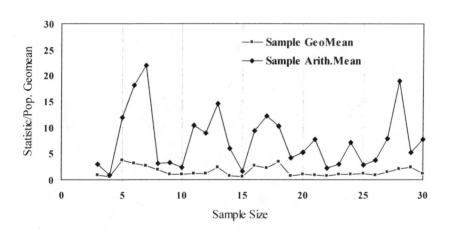

---

[11] A "trial" is a single recalculation of an equation or set of equations with randomly generated variables.
[12] One of the criteria for the selection of the arithmetic mean as a statistical estimator suggested in Gilbert (1987) is a coefficient of variation (η) of less than 1.2. It is noteworthy that approximately one-third of the sample sets meet this criterion (η < 1.2), even though the actual generating function is lognormal.
[13] Note that the mean and variance estimators must be adjusted if sample results below the MDL or PQL are present in the sample data set, as discussed by Berthouex and Brown (1994) and other authors.

The variability of the RME estimates is shown in Figure 3, as a ratio of the population geomean. The SW-846 RMG (the 90% UCL of the log transformed mean) demonstrates good stability for sample sizes greater than 5, and fluctuates around a population geomean ratio of approximately 1.25, without apparent bias. According to SW-846, the 90% UCL of the sample mean should display a predictable relationship to the sample mean, as calculated by the "t" statistic.[14] Figure 3 shows the variation of the sample RME estimates along with the SW-846 "t" statistic, which indicates good agreement with the SW-846 RME, and poor agreement with the RAGs RME.

Figure 3 --

Variation of RME with Sample Size

for Small Data Sets

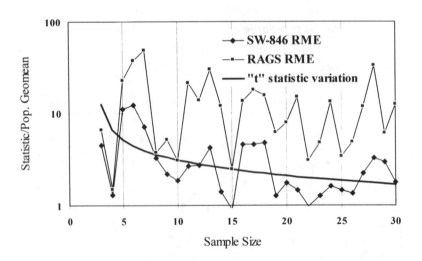

---

[14] The "t" statistic is used as a multiplier for the sample standard deviation divided by the square root of the size of the sample set: $CL \approx t_{\alpha=0.1}[\sigma / \mu] / \sqrt{n}$

There are two independent aspects of the arithmetic statistics of small data sets. Firstly, the arithmetic mean statistic will predict, at best, a multiple (which is approximately 5 for the pdf defined herein) of the sample geomean, which is a stable and conservative estimator of the true average concentration[15]. This introduces a significant overestimation of the exposure concentration, and correspondingly, of the risk to human health due to such contamination. Secondly, the arithmetic statistic will exhibit wide fluctuations around this multiple, such that the population may be overestimated by a factor of 20 or more. This is illustrated in Figure 4. As Figure 5 shows, the overestimation is even more pronounced for the RAGS "Reasonable Maximum Exposure", or RME.[16] The so-called RME over predicts the geomean by a multiple of approximately 10 (Figure 3) for the PDF defined herein, and exhibits fluctuations of up to 60 compared to the population mean. Finally, the variation of the four statistics with sample size was extended to larger data sets of 50, 100, 250, 500, 1000, 2500 and 5000 samples. Figures 6 and 7 show that over 100 samples are needed to control the variability of the arithmetic statistic, whereas the log-transformed statistics have a reasonable variability with as few as 5 to 10 samples.

Figure 4 --
Regression of Sample Means
for Small Data Sets

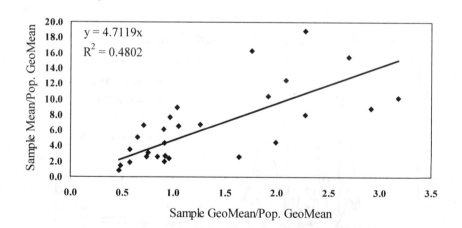

---

[15] Note that in less than 50 of the 2500 trials did the sample geomean underestimate the true average concentration

[16] The MRE is defined as the minimum of the sample maximum and the 95th percentile of the sample arithmetic pdf

Figure 5 --
*Regression of Sample RMEs
for Small Data Sets*

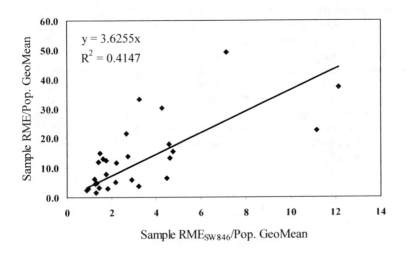

Figure 6 --
Variation of Statistics with Sample Size

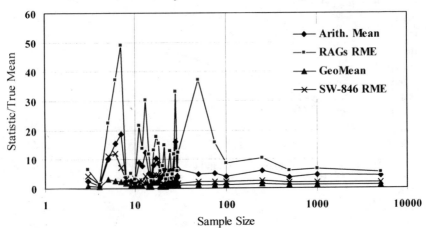

Figure 7 --
Variation of Statistics with Sample Size < 100

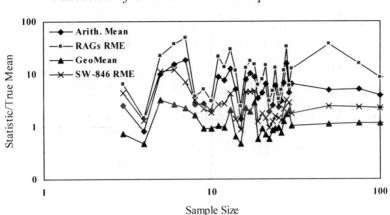

Section IV – Summary

Small data sets are frequently encountered in Land Pollution Control problems. The practitioner needs to calculate a reliable (i.e. with reasonable variance) estimator of the true average concentration of soil and ground water contamination, which is essentially equal to the population geomean. The arithmetic estimators recommended in RAGs are excessively biased and variable, and can result in a 50 to 100-fold overstatement of the true average exposure concentration. The sample geomean and the 90% UCL of the sample geomean are the most appropriate estimators of the average exposure concentration due to their stability for small environmental data sets, and are therefore the most appropriate statistics to compare to risk-based standards.

References

ASTM, 1995, *Guide for Risk-Based Corrective Action Applied at Petroleum Release Sites*, ASTM E 1739-95.

Berthouex, P.M., and Brown, L.C., 1994, *Statistics for Environmental Engineers*, CRC Press, Boca Raton, Florida.

Burmaster, D.E., and Appling, J.W., April 7, 1995, "Introduction to Human Health Risk Assessment, with and Emphasis on Contaminated Properties", *Environment Reporter*, pp. 2431-2440.

Gilbert, R.O., 1987, *Statistical Methods for Environmental Pollution Monitoring*, Van Nostrand Reinhold, New York.

Gilliom, R.J., and Helsel, D.R., 1986, "Estimation of Distributional Parameters for Censored Trace Level Water Quality Data 1. Estimation Techniques", *Water Resources Research*, Vol. 22, No. 2, pp. 135-146.

Gleit, A., 1985, "Estimation for Normal Data Sets with Detection Limits", *Environmental Science and Technology*, Vol. 19, pp. 1201-1206.

Haas, C.N, and Scheff, P.A., 1990, "Estimation of Averages in Truncated Samples", *Environmental Science and Technology*, Vol. 24, No. 6, pp. 912-919.

Helsel, D.R., and Gilliom, R.J., 1986, "Estimation of Distributional Parameters for Censored Trace Level Water Quality Data 2. Verification and Applications", *Water Resources Research*, Vol. 22, No. 2, pp. 147-155.

Newman, M.C., Dixon, P.M., Looney, B.M., and Pinder, J.E. III, "Estimating Mean and Variance for Environmental Samples with Below Detection Limit Observations", Water Resources Bulletin, Vol. 25, No. 4, pp. 905-916.

Ott, W., 1995, *Environmental Statistics and Data Analysis*, CRC Press, Boca Raton, Florida.

Ott, W., 1990, "A Physical Explanation of the Lognormality of Pollutant Concentrations", *Journal of the Air and Waste Management Association*, Vol. 40, pp. 1378-1383.

Shumway, R.H., Azari, A.S., and Johnson, P., 1989, "Estimating Mean Concentrations Under Transformation for Environmental Data with Detection Limits", *Technometrics*, Vol. 31, No. 3, pp. 347-356.

Stoline, M.R., 1991, "An Estimation of the Lognormal and Box and Cox Family of Transformations in Fitting Environmental Data", *Environmetrics*, Vol. 2, No. 1, pp. 85-106.

U.S. EPA, May 1992, *Supplemental Guidance to RAGS: Calculating the Concentration Term*, Publication 9285.7-081.

U.S. EPA, 1986, Test Methods for Evaluating Solid Waste, Field Methods, USEPA SW-846.

Ecological Risk

Thomas P. Burns[1], Barney W. Cornaby[1], Stephen V. Mitz[1], and Charles T. Hadden[1]

A PROBABILISTIC INTERPRETATION OF THE QUOTIENT METHOD FOR CHARACTERIZING AND MANAGING RISK TO ECOLOGICAL RECEPTORS

REFERENCE: Burns, T. P., Cornaby, B. W., Mitz, S. V., and Hadden, C. T., "**A Probabilistic Interpretation of the Quotient Method for Characterizing and Managing Risk to Ecological Receptors,**" *Superfund Risk Assessment in Soil Contamination Studies: Third Volume, ASTM STP 1338*, K. B. Hoddinott, Ed., American Society for Testing and Materials, 1998.

ABSTRACT: Hazard or 'risk' quotients are commonly used to characterize the risk to ecological receptors at Superfund sites. A risk quotient is the ratio of an estimated exposure level (concentration or dose) to an effects threshold level (concentration or dose). Ecological hazard quotients can be formulated and interpreted in probabilistic terms. The numerator of the ratio is the exposure corresponding to the maximum allowable risk to the receptor at the site. Given a distribution of exposures at a site, the specified exposure level defines the maximum acceptable probability that a receptor experiences unacceptable harm, i.e., maximum acceptable risk. The exposure level at Superfund sites is frequently specified by regulators to be that corresponding to the upper 95th confidence limit on the mean concentration. The denominator of the ratio is the exposure corresponding to the dose causing the maximum allowable adverse effect on the receptor. Given a distribution of exposures at a site, the effects threshold defines the actual probability that the receptor is exposed to levels greater than the threshold, i.e., actual risk. The effects threshold at Superfund sites is frequently required by regulators to be that corresponding to the no effect level. This formulation of ecological hazard quotients clarifies the meaning of the quotient and highlights the importance of the exposure level and effects thresholds specified by regulators. This paper also discusses the relationship between this probabilistic interpretation of risk quotients and probabilistic exposure modeling, e.g., Monte Carlo simulation.

---

[1] Engineering and Environmental Management Group, Science Applications International Corporation, P.O. Box 2502, 800 Oak Ridge Turnpike, Oak Ridge, TN 37831-2502

KEYWORDS: Ecological risk assessment, hazard quotient, Monte Carlo methods, probability theory, risk-based decision making, risk management, uncertainty

---

Risk is the probability of experiencing an injury or loss, i.e., being harmed. Ecological risk assessments for Superfund sites must evaluate the probability or likelihood that non-human organisms (ecological receptors) are being or will be harmed by exposure to contaminants present in soil or other environmental media (EPA 1992). Ecological receptors are individual organisms with special status (e.g., T&E species) or populations of "common" organisms potentially exposed to contaminants at the contaminated site.

There is no single correct or established approach for assessing the risk to ecological receptors at Superfund sites with soil contamination. A desirable approach is to perform controlled experiments to determine if the response of a test organism exposed to soil from the site is significantly different from the response of test organisms exposed to uncontaminated control soils (Wentsel et al. 1996). The risk to ecological receptors can be assessed indirectly when resources do not allow or circumstances do not warrant the direct toxicity testing of the soil.

The most often used quantitative indirect approach to assessing risk to a receptor from a soil contaminant is commonly referred to as the quotient method (Barnthouse et al. 1986, EPA 1992). In this approach, one compares the measured or predicted concentration or dose of the contaminant to which the receptor is exposed (the exposure level) to a concentration or dose known or thought to be associated with a specified response in the receptor organisms or other test organisms (the effect level). The ratio of the two concentrations or doses is referred to as a hazard or risk quotient. A ratio greater than or equal to 1 is typically interpreted as meaning that the receptor is at risk from the contaminant. Conversely, if the ratio is less than 1, it is usually concluded that the receptor is not at risk.

It is reasonable to ask, "If risk is a probability, how can the ratio of an exposure and an effect level indicate whether there is or is not risk?" This paper presents a way to understand risk quotients in terms of probabilities. The purpose of this paper is not to discuss pros and cons of different approaches, nor defend the quotient method over other more direct approaches. With a better understanding of risk quotients, we can better apply the quotient method when it is the preferred or default approach.

The Probability of Dose-Response Relationships

The degree of harm experienced by an individual when it is exposed to a dose of contaminant can take various values. The degree of harm (Y) can thus be thought of as a random variable. A random variable (Y) can take different values with varying probabilities $P(Y)$, such that $\sum_y P\{Y=y\} = 1$ (Ross 1984).

The probability that an individual organism will experience a given degree of harm as a result of exposure to a contaminant is conditional on the magnitude of its exposure to the contaminant. The evidence for this proposition comes from the numerous experiments exposing different individuals of a given type to fixed concentrations or doses of a toxicant. These experiments ideally find that, for concentrations between the largest causing no effect on every individual (EC_0) to the smallest causing effect on all individuals (EC_{100}), different degrees of harm are experienced by those individuals (Fig 1). For example, the well-known toxicity benchmark referred to as the LD_{50} is defined as the dose that is lethal to 50% of the individuals exposed to that concentration. Thus, in the case where the degree of harm resulting from exposure (Y), is conditional upon the magnitude of exposure (X), we have that $P\{Y=y|X=x\}$, where $\Sigma_y\ P\{Y=y|X=x\} = 1$.

Depending on what kinds of effect are being observed, the degree of harm may be a continuous variable (e.g., growth) or a discrete variable (e.g., death). For the purpose of this paper and without loss of generality, we deal only with the continuous variable case. Further we assume that the degree of harm Y must assume some nonnegative value for every exposure X between $-\infty$ and $+\infty$. These assumptions allow the application of the rules of probability to the dose-response relationship.

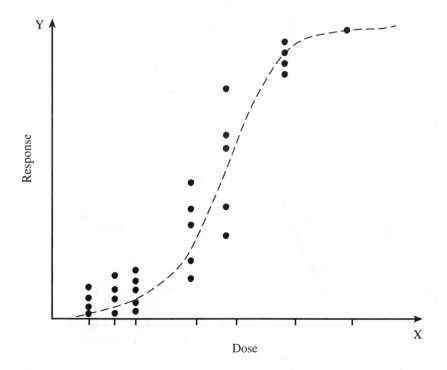

FIG. 1 Hypothetical results of toxicity tests showing variation in response to a given dose.

From the definition of a cumulative distribution function of a random variable (Ross 1984), if there is defined a maximum allowable (threshold) effect, T_y, then the probability that the degree of harm experienced by an individual when exposed to a given dose of contaminant ($X=x$) is less than or equal to T_y is

$$F(T_y)_x = P\{Y \le T_y | X=x\}. \tag{1}$$

to the contaminant (Fig. 2). One can assume that the distribution function $F_x(y)$ has a particular shape, or $F_x(y)$ can be determined by a series of toxicity experiments, as suggested above. For the purposes of this paper, there is no need to specify the shape of the probability density function or cumulative distribution function.

In the practice of risk assessment at Superfund sites, the variation in the response to a given dose is often ignored. A single dose or concentration associated with an acceptable degree of adverse effect is identified and used as the effects threshold, X_T. This is typically the case with the use of the quotient method in ecological risk assessment. Using a single-valued effects threshold for a given receptor and contaminant is unlikely to underestimate risk if that effects threshold corresponds to

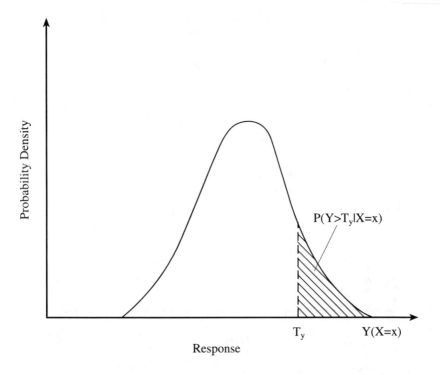

FIG. 2 Hypothetical distribution of responses among individuals exposed to a given concentration or dose of contaminant ($X=x$). The cross-hatched region defines the probability that an individual response exceeds the maximum acceptable response (T_y).

the no effect level. The benefit of defining a single effects-threshold is that it is easy to understand and it allows for a point estimate of the actual risk to the receptor at the site.

The Probability of Exposure

At any given Superfund site, the magnitude of individual exposures to a soil contaminant is also likely to be a random variable. In general, contaminant concentrations in soil or other environmental media vary, as do the experiences of individual organisms. Organisms have different experiences as a result of their different behaviors, physiologies, locations, movements, etc. Contaminant concentrations at a site vary for many reasons, including fate and transport mechanisms and heterogeneous biochemical or physical degradation rates. Because of this, the single actual magnitude of exposure experienced by each individual organism cannot be predicted in advance. Furthermore, an individual's actual exposure is only one of a bounded range of possible magnitudes that the exposure might take. Thus, the exposure that an individual organism experiences at a site can be modeled as a random variable (Fig. 3). That is, the probability that a given individual experiences exposure $X = x$ is $P\{X=x\}$, where $\Sigma_x P\{X=x\} = 1$.

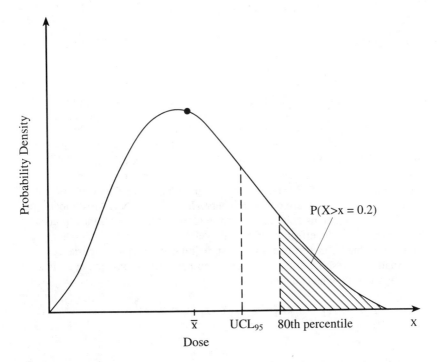

FIG. 3   Hypothetical distribution of exposures among individuals at a contaminated site showing the mean, upper 95th confidence limit on mean and 80th percentile values.

If the exposure at a site is a random variable with a distribution of values, then specifying a dose defines the probability of experiencing a dose greater than that specified. When the specified dose is the dose associated with the maximum allowable effect, then that dose defines the probability of experiencing unacceptable harm, i.e., risk.

The Probability of Risk

Assuming no avoidance behavior of any sort, the probability that an individual organism experiences a dose of a given magnitude, $P\{X=x\}$, is independent of the probability of experiencing a particular degree of harm given that it experiences that dose, $P\{Y=y|X=x\}$. This is a somewhat simplistic model because it assumes that organisms do not become progressively intoxicated resulting in changes in behavior or physiology affecting subsequent exposure. Note that, in this simple model, the exposure experienced is independent of the resulting effect, although the effect is not independent of the exposure. The exposure is the cause on which the effect is conditional.

The causal relationship between exposure (cause) and toxicity (effect) is captured in Baye's Formula,

$$P(E|C) = P(EC)/P(C),$$

which by rearranging gives,

$$P(EC) = P(E|C)P(C).$$

Baye's Formula states that the probability that a conditional event (E), here the 'effect', and the event on which it is conditioned (C), here the 'cause', both occur is equal to the product of the probabilities that the cause occurs, $P(C)$, and the probability that the effect occurs given that the cause occurs, $P(E|C)$. When there are multiple possible conditioning events, C_i, $i = 1, 2, ..., $ n, then Baye's Formula takes the sum of the products over all i. Note the similarity of Baye's Formula to the definition of independent events, $P(FG) = P(F)P(G)$. In Baye's Formula, one of the events is conditional on the other, so they are not both independent of the other. Most models of causality in western civilizations hold that the cause (conditioning event) must precede and thus is independent of the effect (conditioned event).

By Baye's Formula, the probability that an individual organism experiences an effect greater than the threshold effect, T_y, from exposure to a contaminant at a given site is defined by the following equation,

$$P\{Y > T_y\} = \Sigma_x P\{Y > T_y|X=x\} \times P\{X=x\}, \tag{2a}$$

that is,

$$P\{Y > T_y\} = \Sigma_x [(1 Ð P\{Y \leq T_y|X=x\}) \times P\{X=x\}] \tag{2b}$$

$$= 1 - \Sigma_x \left(P\{Y \leq T_y | X=x\} \times P\{X=x\}\right), \qquad (2c)$$

because $\Sigma_x P\{X=x\} = 1$ at any given site. Equation 2c expresses the risk as the complement of the probability that the individual does not experience an effect greater than the maximum allowable effect, T_y. In most cases exposure is a continuous variable, so the summation in (2) is a discrete approximation of the integral.

Equation 2 expresses risk as a function of the probability of exposure and the conditional probability of experiencing a degree of harm greater (2a) or less (2b, 2c) than a maximum acceptable degree of harm, T_y. If the threshold effect, T_y, is realized at all exposures greater than a single threshold exposure, then at any given site

$$P\{Y > T_y\} = P\{X > X_T\}, \qquad (2d)$$

$$= 1 - P\{X \leq X_T\} \qquad (2e)$$

where X_T is the largest exposure not associated with (i.e., causing) harm greater than the maximum allowable effect.

Risk managers, with input from ecologists, must define T_y, the maximum allowable effect, and P_{max}, the maximum acceptable risk, that is, the maximum acceptable probability of an individual experiencing an unacceptable harm. Setting those parameters cannot be solely the responsibility of risk assessors because doing so requires value judgments about what should be protected and to what degree (EPA 1992).

Equation 2 is the risk to the average individual, i.e., the probability that an individual will experience an unacceptable degree of harm The risk to receptor populations is discussed below.

The risk to populations can be thought of as the probability that some fraction of individuals in the receptor population experience an effect greater than some maximum allowable effect, T_z. The number of individuals that experience an independent event that occurs with a probability p can be modeled as a binomial random variable with parameters p and n, where n is the number of trials (Ross 1985). In the case of a population at a contaminated site, n is the number of individuals in the exposed population. Assuming that individuals' exposures are independent, the expected number of individuals that will experience an event that occurs independently with probability p, given n individuals in the population, is then simply np, and the expected fraction of individuals experiencing the event is $f = np/n = p$, where p is the probability of the event occurring to any one individual. The probability of an individual in the receptor population experiencing an effect greater than T_z given that it is exposed to contaminant concentration $X=x$ over all x was shown in Eq. (2) to be $P\{Y>T_z\}$. Thus,

$$f = P\{Y > T_z\}. \qquad (3)$$

If individuals are not distributed randomly with respect to each other such as is the case with aggregation or dispersion, then the binomial model may not apply and f will in general not be equal to $P\{Y > T_z\}$, the probability that an individual experiences harm, T_z.

This simple model is sufficient to show that, in principle, the risk to populations can also be expressed in terms of the maximum allowable degree of harm to individuals in the population (T_z) and the fraction of individuals that experience this level of effect, f. As with the model for risk to individuals, risk managers have the responsibility to define both T_z, the maximum allowable effect on individuals in the population, and f_{max}, the maximum acceptable fraction of individuals in the receptor population experiencing an effect greater than T_z.

Where both populations of common species and individuals of species with special status are potentially exposed to contaminants at a site, the risk managers must specify the risk parameters for both. The acceptable risk to populations, $f_{max} = P\{Y > T_z\}$, may not equal the acceptable risk to special status individuals, $P_{max} = P\{Y > T_y\}$, and T_y for special status individuals may not be the same as T_z for individuals of species without special status. For example, the maximum allowable effect for individuals of a T&E species may be no effect, and P_{max} may be zero or some other minimal probability, e.g., 0.01. For species without special status, the maximum allowable effect for any given individual may be as severe as death, and f_{max} may be much higher than that for T&E species, e.g., 0.20. The formulation of risk described above allows for this flexibility and thus, a simple approach to ecological risk assessment using risk quotients.

Interpreting Ecological Hazard Quotients

A risk quotient for an ecological receptor is typically the ratio of a predicted exposure level (concentration or dose) and a "toxicity benchmark" or effects-threshold level. The exposure value used as the numerator to calculate the risk quotient is usually the expected (mean, average) exposure level or a conservative estimate of the mean, e.g., 95th percentile upper confidence limit on the mean (UCL$_{95}$). The numerator could just as well be the median, mode, maximum, or any other parameter of the distribution of exposure values in the population, e.g., the 80th percentile. At Superfund sites, stakeholders with regulatory authority frequently require that the exposure point concentration be the lesser of the UCL$_{95}$ and the maximum. The effects-threshold level used in the denominator of the quotient is usually the test concentration or dose producing a particular level of effect, e.g., the no observed adverse effect level (NOAEL) or lowest observed adverse effect level (LOAEL). At its simplest, assuming no uncertainty about the true exposure and effects-threshold levels, the risk quotient indicates whether there is risk. That is, if the ratio exceeds unity then there is a risk problem for that receptor. The problem for ecological receptors if risk quotients exceed unity can be explained in terms of the probabilities discussed above.

The probabilistic interpretation of risk quotients begins with the distribution of exposures at a site (Fig. 3). For any given distribution of exposures, a given effects-threshold, $X(T_z)$, defines the fraction of the individuals at the site that is exposed to a concentration greater or equal to the effects threshold concentration, f (Fig. 4). That is, $X(T_z)$ defines the actual risk at the given site, f, and $1-f$ is the fraction that is exposed to less than the effects threshold concentration.

If f, the actual risk, exceeds f_{max} , the maximum allowable risk as defined by the risk manager, then there is by definition a problem. For example, if $f_{max} = 0.2$, then the exposure corresponding to the effects-threshold, $X(T_z)$, must be greater than the exposure corresponding to the 80th percentile on the distribution ($1 - f_{max}$), i.e., the exposure above which 20 percent of the individuals are exposed. If the effects-threshold, $X(T_z)$, corresponds to the 70th percentile exposure in the population, for example, then f equals 0.3, which exceeds f_{max}, the maximum allowable fraction of individuals exposed to concentrations exceeding the effects-threshold concentration. That is, 10% too many individuals are likely to experience the maximum effect or greater, and thus there is unacceptable risk to the population. In terms of doses, the dose corresponding to the maximum acceptable risk exceeds the dose corresponding to the maximum allowable adverse effect, i.e., the maximum allowable dose.

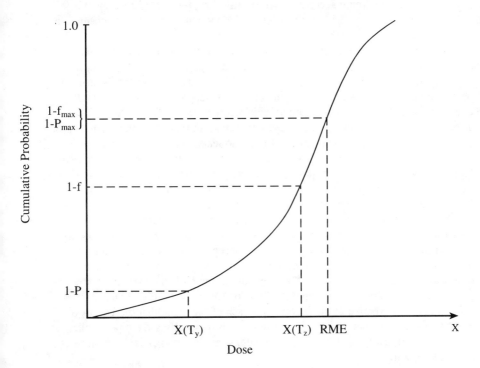

FIG. 4   Hypothetical cumulative probability function for exposures to a soil contaminant at a site, showing relationship between f_{max} and P_{max}, the maximum allowable risk as determined by the value at the site of the specified exposure parameter, the reasonable maximum exposure (RME), and the actual cumulative probability of being exposed to greater or equal to the effects threshold for populations and individuals, respectively, $X(T_z)$ and $X(T_y)$. The cases depicted indicate unacceptable risk because the actual probabilities of exposure above the effect thresholds exceed the maximum allowable probabilities.

Likewise, any given individual will experience an exposure greater than the dose associated with the effect-threshold (presumably the dose causing the specified maximum allowable effect, T_y) with a probability specified by the distribution of exposures. This probability can be denoted P and $1-P$ is the probability that an individual will not experience the maximum allowable adverse effect or greater. If P exceeds $P_{max} = P\{Y > T_y\}$ as defined by the risk manager, then there is a problem for individual receptors. For instance, if $P_{max} = 0.05$, then the effects threshold, $X(T_y)$, must be greater than the exposure equal to the 95th percentile of the exposure distribution at the site. If the effects threshold falls at the 88th percentile value for the site, then P is equal to 0.12, which exceeds 0.05. For this case as the one before, the problem is that the probability of an individual in the receptor population experiencing an effect greater than T given that it is exposed to contaminant concentration $X=x$ over all x, is greater than maximum acceptable probability, i.e., the risk is unacceptably large. As in the earlier case, the interpretation of the quotient in terms of exposures (or doses) tends to mask the interpretation in terms of risk.

This interpretation of risk quotients demonstrates that the exposure distribution parameter used as the quotient numerator (e.g., the 80th percentile, UCL_{95}) and the effects threshold used as the denominator (e.g., NOAEL, LOAEL) define the critical risk parameters. The numerator specifies the maximum acceptable risk, f_{max} or P_{max}. The denominator specifies the maximum acceptable effect, T_y or T_z. These risk parameters deserve serious attention because important risk management decisions are being made on the basis of the magnitude of risk quotients.

The probabilistic interpretation of risk quotients clarifies the inconsistency of using a statistic like the UCL_{95} for the exposure parameter (numerator) and motivates using a fixed percentile value. Using a statistic such as the UCL_{95} or maximum detected concentration as the specified exposure leads to potentially varying degrees of protection for different contaminants and locations, unless all contaminants are everywhere distributed identically. This is unlikely. At any given site, the UCL_{95} for one contaminant might correspond to a maximum acceptable risk of 0.35 and for another 0.21. By specifying a percentile of the distributed exposures as the specified exposure used in the numerator, e.g., 80th percentile, one fixes the maximum acceptable risk across all contaminants regardless of how they are distributed. Therefore a better alternative to the current approach of requiring that the UCL_{95} or maximum be used in the numerator of risk quotients is to specify a fixed percentile based on ecological theory or data about populations' abilities to persist in the face of environmental insult.

For the probabilistic interpretation of risk quotients to make sense, the effect associated with the effects threshold in the denominator must be the maximum allowable effect, T_z or T_y. For example, the LOAEL estimated by a toxicity test is based on a statistically significant difference between the fraction of individuals experiencing some adverse effect as a result of exposure to the LOAEL dose and the fraction of individuals in the control population experiencing the same adverse effect. If the empirically determined exposure corresponding to the specified exposure parameter exceeds an effects threshold corresponding to the LOAEL dose, then the fraction of the population at the site exposed to a dose higher than the LOAEL will be greater than the fraction of the test population that experienced the effect. If this effect was not the maximum allowable effect, then it would not matter that the fraction at the site exceeded the fraction in the test. This would contradict the usual meaning of a risk quotient greater than one.

Given the central importance of the risk parameters to the meaning of risk quotients, it is appropriate that risk managers have primary responsibility for specifying those parameters. Regulators have not as yet explicitly identified the effect associated with the specified effects threshold as the maximum allowable effect for ecological receptors. Doing so would generate much needed debate and perhaps the research needed to make a sound determination of what is a serious ecological effect. Nor has risk assessment guidance been promulgated specifying a maximum acceptable risk for protected individuals or populations of organisms without special status.

Discussion

Uncertainty about the true value of the specified exposure parameter value (e.g., the 80th percentile) and effects values (T_y) clouds but does not eliminate the probabilistic interpretation of hazard quotients. Discrepancy between the actual properties of something and our observations or measurements of it, i.e., error, is a problem shared by all risk assessment methods. Uncertainty about these values also comes from natural variability, which makes a probabilistic interpretation necessary. There would be little need to talk of risk or probability of harm if there was no variability in nature or our ability to make measurements. The solution is to acknowledge the uncertainty in the exposure and effects values and deal with it.

There are two simple ways to deal with the uncertainty of hazard quotients. Both take advantage of the fact that there is a single decision criteria in the quotient method, that is, the relationship of the quotient to the threshold value of 1. One approach is to define a range of values which are treated as being equivalent. For example, quotients ranging in value from 0.8 to 1.2 can be treated as being equal to 1 and interpreted identically. The more common approach is to consistently use conservative estimates of the exposure and effects values. The resulting quotient overestimates the true ratio of the actual exposure and effects values. Both these approaches usually suffer from not having quantitative estimates of the uncertainty in the exposure and effects values and the resulting risk quotient.

The probabilistic interpretation of hazard quotients is intimately related to a third approach to dealing with uncertainty — probabilistic uncertainty analysis. Methods such as Monte Carlo analysis usually generate a distribution of exposure values, sometimes a distribution of effects thresholds. If the risk assessor has been given only a fixed effects threshold associated with a maximum allowable degree of harm $X(T_z)$ or $X(T_y)$, then the resulting output explicitly defines a risk quotient if the risk manager also specifies an acceptable fraction of the modeled exposures that can equal or exceed the effects threshold, i.e., f_{max} or P_{max}. Probabilistic methods offer the additional advantage to ecological risk assessors of being able to quantify the effect of multiple sources of uncertainty in the exposure or effects estimates on the resulting estimate of risk

(Hoffman and Hammands 1994). But if risk managers do not specify a maximum allowable effect (T_y or T_x) and a maximum allowable probability that an individual experiences this effect, f_{max} or P_{max}, then such advanced probabilistic modeling and analysis methods offer no advantage for ecological risk assessment.

Probabilistic methods of assessing risk are gaining greater acceptance among risk assessors and regulators. Risk quotients have been shown to be a simple expression of the same probabilistic exposure model these methods simulate. Both the quotient method and more detailed probabilistic models, e.g., Monte Carlo analysis, require stakeholders to specify the risk parameters, T_y or T_z and f_{max} or P_{max}. Given specified risk parameters and a distribution of contaminant concentrations from the site, either a single quotient or a distribution of quotients generated by probabilistic models will be available to characterize the probability of harm to ecological receptors.

REFERENCES

Barnthouse, L.W., Suter, G.W., Bartell, S.M., Beauchamp, J.J., Gardner, R.H., Linder, E., O'Neill, R.V., and Rosen, A.E., 1986. Users Manual for Ecological Risk Assessment. Publ. No. 2679, ORNL-6251, Environmental Sciences Division, Oak Ridge National Laboratory, Oak Ridge, Tn.

EPA, 1992, Framework for Ecological Risk Assessment, EPA/630/R- 92/001, Risk Assessment Forum, U.S. Environmental Protection Agency, Washington, D.C.

Hoffmann, F.O. and Hammands, J.S., 1994, "Propagation of Uncertainty in Risk Assessments: The Need to Distinguish Between Uncertainty Due to Lack of Knowledge and Uncertainty Due to Variability", Risk Analysis, Vol. 14, pp. 707Ð717.

Ross, S. 1984. A First Course in Probability. Second Edit. Macmillian Publishing Co., NY.

Ross, S. 1985. Introduction to Probability Models. Academic Press, Inc., Orlando, Fl.

Wentsel, R.S., La Point, T.W., Simini, M., Checkai, R.T., Ludwig, D., and L. W. Brewer 1996. Tri-Service Procedural Guidelines for Ecological Risk Assessments, ERDEC-TR-221, Edgewood Research Development & Engineering Center, Aberdeen Proving Ground, MD.

Greg Linder,[1] Michael Bollman,[2] Clarence Callahan,[3] Christopher Gillette,[4] Alan Nebeker,[5] and David Wilborn[6]

BIOACCUMULATION AND FOOD-CHAIN ANALYSIS FOR EVALUATING ECOLOGICAL RISKS IN TERRESTRIAL AND WETLAND HABITATS:

AVAILABILITY-TRANSFER FACTORS (ATFs) IN 'SOIL → SOIL MACROINVERTEBRATE → AMPHIBIAN' FOOD CHAINS

REFERENCE: Linder, G., Bollman, M., Callahan, C., Gillette, C., Nebeker, A., and Wilborn, D., **"Bioaccumulation and Food-Chain Analysis for Evaluating Ecological Risks in Terrestrial and Wetland Habitats: Availability-Transfer Factors (ATFs) in 'Soil → Soil Macroinvertebrate → Amphibian' Food Chains,"** *Superfund Risk Assessment in Soil Contamination Studies: Third Volume, ASTM STP 1338*, Keith Hoddinott, Ed., American Society for Testing and Materials, 1998.

ABSTRACT: As part of the ecological risk assessment process for terrestrial and wetland habitats, the evaluation of bioaccumulative chemicals of concern (BCCs) is frequently pursued through food-chain analysis with a subsequent comparison of daily doses to benchmark toxicity reference values, when available. Food-chain analysis has frequently been applied to the analysis of exposure to BCCs identified as chemicals of potential ecological concern (COPECs) in the ecological risk assessment process. Here, designed studies focused on wetland food-chains such as "hydric soil → soil macroinvertebrate → amphibian" and terrestrial food-chains such as "soil → plant → small mammal" illustrate an approach for the derivation and validation of trophic transfer factors for metals considered as COPECs such as cadmium, chromium, copper, lead, and zinc. The results clearly indicate that the transfer of chemicals between trophic levels is critical in the bioaccumulation process in wetland and terrestrial food-chains and is influenced by numerous interacting abiotic and biotic factors, including physicochemical properties of soil, and the role, if any, that the metal has in the receptor as a required trace element.

[1] Applied Ecologist, HeronWorks Farm, 5400 Tacoma Street NE, Brooks OR 97305.
[2] Botanist, Dynamac International, Inc., 200 SW 35th Street, Corvallis OR 97331.
[3] Ecologist, US EPA, Region 9, 75 Hawthorne Street, San Francisco CA 94105.
[4] Graduate Student, Department of Microbiology, Oregon State University, Corvallis OR 97333.
[5] Research Scientist, US EPA, Environmental Research Laboratory, 200 SW 35th Street, Corvallis OR 97331.
[6] Applied Biologist, Takena Ecological Services, Corvallis OR 97333.

KEYWORDS: food-chain analysis, ecological risk assessment, wildlife, soils, metals

---

Food-chain analysis has frequently been applied to the analysis of exposure to bioaccumulative chemicals of concern (BCCs) or chemicals whose cumulative dose may be associated with adverse biological effects. Regardless of their potential toxicity, such chemicals may be identified as chemicals of potential ecological concern (COPECs) in the ecological risk assessment process. In general, all food-chain models follow this structure:

$$IR_{chemical} = \sum [(C_i \times M_i \times ATF_i)/BW]$$

where

$IR_{chemical}$	=	species-specific total rate of intake of chemical by ingestion (mg/kg-day), or dose;
C_i	=	chemical concentration in medium i (mg/kg in soil, μg/L in water, and various dietary constituents in mg/kg in food);
M_i	=	rate of ingestion of medium i (e.g., kg/day and μg/L);
ATF_i	=	availability-transfer factor for chemical in medium i; and
BW	=	body weight of receptor species (kg).

Although bioaccumulation is conceptually similar in both terrestrial and aquatic habitats, the role of bioaccumulation and food-chain analysis in ecological risk assessment differs across these habitats. These differences stem in part from matrix effects that directly or indirectly influence the extent and rate of bioaccumulation in various habitat settings.

For example, in the general food-chain model, ATF_i represents a complex transfer function for a chemical from one material in a food-chain to another (e.g., transfer of metal in soil to vegetation, or transfer of metal residues in vegetation to herbivore). ATF_i is highly matrix-dependent and variable, especially across soil types ranging from upland soils to wetland soils; ATF_i also reflects biological mechanisms, e.g., physiological and biochemical, that influence uptake and assimilation of chemicals from soils. Hence, ATF_i is a source of much uncertainty in ecological risk assessments. However, by using standardized ASTM bioaccumulation tests conducted in parallel with food-chain modeling efforts, ratio estimators of ATF_i (e.g., biota:soil ratios) can be derived. Similarly, other values critical to food-chain analysis such as ingestion rates can be described. The present study focuses on wetland food-chains such as "hydric soil → soil macroinvertebrate → amphibian" to illustrate an approach for the derivation of ATFs for metals in soil, including cadmium, chromium, copper, lead, and zinc. A similar analysis of terrestrial food-chains such as "soil → plant → small mammal" will be summarized in a future paper (Linder et al., in preparation).

Methods and Materials

Food-Chain Analysis in Feeding Trials Using Amphibians

As noted in the introduction, all food-chain models are designed as:

$$IR_{chemical} = [(C_i * M_i * ATF_i)/BW]$$

which, in this screening-level effort (Linder et al., 1993) was implemented as a deterministic model summarized in Figure 1.

Given the designed laboratory feeding studies that supported the implementation of the food-chain analysis, dose equations for amphibians in these studies simplified to food consumption rate (FR) times concentration of metals in earthworms (C_{food}), including coincidental ingestion of soils retained in the gut. Measured concentrations of COPECs soils and tissues -- in these studies, cadmium, chromium, copper, lead, and zinc -- were used throughout the analysis.

Since metal concentrations in earthworms fed to frogs were known from this study, ATFs were not required to estimate concentrations of COPECs in feed based upon soil concentrations. ATFs, however, have been calculated for subsequent application to future food-chain analyses that lack concentration terms for metal residues in feed. In this controlled laboratory study, exposure to metals was restricted to sources in feed stocks; that is, other routes of exposure contributed little, if any, COPEC to the exposure.

Laboratory Feeding Study: Bioaccumulation of Metals in Earthworms and Trophic Transfer to Amphibians via Feeding Exposure

In the laboratory, bioaccumulation of metals was evaluated using a representative soil invertebrate, the compost worm *Eisenia foetida* (ASTM 1998). Earthworms were exposed to contaminated soils and reference soils. Reference soils were used "neat" as a laboratory control as well as being used as a soil amendment; that is, metal-contaminated soils were mixed with reference soils in a Hobart mixer to yield a dilution series of 100%, 50%, 25%, 12.5%, and 0% contaminated soil (dry weight basis; EPA 1989). Earthworms were then exposed to these soils for 28 days and were subsequently used as feed for frogs previously experienced at eating live-feed earthworms.

In these controlled feeding trials, the African Clawed Frog (*Xenopus laevis*) was used as the animal model. The species has been used in numerous related studies in support of ecological risk assessments for wetland habitats, in preliminary studies focused on determining ingestion rates (e.g., *Eisenia* consumption/frog/day), and in lead accumulation studies using a diet of earthworms collected from highly polluted and relatively unpolluted soils (Ireland 1977). For this test effort, juvenile frogs were exposed to different metal-contaminated earthworm diets.

Exposure of amphibians through metal-contaminated worms included 10 replicates each of four concentrations plus controls. One laboratory control included worms held in

$$\text{Ingested dose} = \frac{[[C_{food(s)} \text{ (ug/gm)} \times FR \text{ (gm/day)}] + [C_{med}\text{(ug/gm)} \times SC \text{ (gm/day)}] + [C_{water}\text{(ug/ml)} \times WC \text{ (ml/day)}]] \times ATF}{BW \text{ (kg)}}$$

where

$C_{food(s)}$	=	concentration of chemical in food items (plant and animal food items);
FR	=	foraging rates for plant and animal materials;
C_{med}	=	concentration of chemical in environmental media (sediment and soil);
SC	=	intentional or coincidental rate of soil and sediment consumption;
C_{water}	=	concentration of chemical in drinking water;
WC	=	water consumption rate
ATF	=	availability-transfer factor for chemical of concern in soil; and
BW	=	body weight of receptor species

Note: $C_{food(s)}$ may be decomposed into as many categories as necessary; that is, $[(C_{food,1}) \times FR_1 + (C_{food,2}) \times FR_2 + \ldots + (C_{food,n}) \times FR_n]$ to account for all foraging sources.

FIG. 1 -- *Summary of determinist food-chain model which was implemented in the amphibian feeding trials.*

uncontaminated soil (reference control), while an additional control included worms held in vermiculture media or artificial soil (ASTM 1998). During feeding trials, mortality and behavioral observations were recorded. Ten 3 in. to 4 in. juvenile frogs were exposed at each metal-contaminated earthworm diet, for a total of 50 frogs for the study. Males and females were assigned randomly to treatments, with each treatment receiving the same approximate ratio of males to females.

Frogs were held individually in 1.7-L exposure chambers. Continuous-flow well water (25°C) was dispensed to exposure vessels through drip irrigation emitters inserted into the lids of the exposure chambers. The exposure chambers were then placed in a flow-through stainless steel water bath also maintained at 25°C, and an outlet located 4-cm above the surface of the water bath on the sidewall of the exposure chamber assured each chamber would be maintained at 25°C throughout the feeding trial. Temperature in the water bath was monitored using a circular chart recorder.

For amphibians, preliminary studies using cadmium chloride suggested that acute effects might be expected only at dietary concentrations approaching those which were obtained in worms exposed at the 100% soil concentration. Unpurged worms were used in this feeding study to reflect a realistic food consumption scenario; based on past experience, gut content contributed about one-half the metal load in unpurged earthworms (on a dry weight basis). Bioaccumulation determinations with respect to earthworms (that is, determination of tissue residues in purged individuals) are planned for subsequent publication (Linder et al., in preparation). Different diet concentrations were obtained by exposing earthworms directly in soil mixtures (12.5, 25, 50, and 100%) to obtain site-specific relationships between metal concentrations in soil and metal concentrations in earthworms. Forty-five worms were introduced into 400 g wet weight soil mixtures twice weekly for 4 weeks in series. Each jar of earthworms was exposed for 28 days.

Live earthworms were fed to frogs at a rate of 3 adult clitellate worms per animal twice weekly. For feeding, earthworms were collected from exposure jars, rinsed clean of soil with reagent grade water, weighed in groups of 3, and hand fed to frogs. Each worm group weight was recorded for each frog immediately prior to feeding. A rubber stopper was placed in the drainage hole of each exposure vessel immediately prior to feeding, and the flow-though water shunted directly into the bath. After 24 hours, each vessel was checked, any remaining worms were collected, weighed, and subtracted from the offered food weight. In general, frogs ate all earthworms offered in the feeding trial. Following each feeding and collection of uneaten food, the stopper was then removed from the exit hole, and the flow-through water restored to the vessels. The total weight of ingested worms was calculated for each frog for the study duration. When present, the uneaten earthworms from each daily jar were rinsed clean with reagent grade water, frozen, and held until the end of the feeding study. All daily worm groups for each concentration were analyzed via inductively coupled plasma-atomic emission spectroscopy (ICP-AES), and total ingested cadmium, copper, chromium, lead, and zinc was calculated for each frog.

At the termination of the study, samples of thigh (bone included), liver, and eggs (when available) were collected from frogs in each treatment and analyzed via ICP-AES.

Soil Analysis

Subsamples of soils were analyzed for particle size distribution (pipette method), total organic carbon, cation exchange capacity, exchangeable bases, salt pH and other parameters (Klute 1986).

Results

Concentrations of Metals in Soil and Metal Uptake in Earthworms

Soils used in these studies were characterized as loams to sandy loams, with pH ranging from 5.1 to 6.2 and cation exchange capacity from 16.1 to 21.8 mEq/100g. At collection, the soils were not saturated, but seasonal ponding in the wetlands from which samples were collected yielded chroma and hue consistent with wetland soils (Lyon 1993).

Table 1 presents metal concentrations in the soil mixtures prepared with site soil and reference soil, as well as for artificial soil. Twenty-eight day laboratory bioaccumulation studies completed with these soils showed no mortality in any treatment, even at 100% soil concentration. At the 100% contaminated soil concentration, cadmium and chromium concentrations were elevated relative to the reference and artificial soil, and occurred at concentrations that had been associated with adverse effects in soil biota in other studies (Alloway 1990). Zinc and copper concentrations were also elevated relative to reference and artificial soils, but the concentrations did not appear sufficient to be associated with adverse effects based only on their concentrations. Lead concentrations were only slightly elevated over reference and artificial soils. The 50%, 25% and 12.5% mixtures of soil presented metal concentrations that were correspondingly less than metal concentrations in neat soils following their amendment with reference soil. All soil concentrations were determined from earthworm exposure jars at the time earthworms were collected for feeding to frogs.

Concentrations of metals in earthworms are presented in Table 2. Each treatment is reported as mean $\pm$ standard deviation (n = 10). Of the metal concentrations in earthworms fed to frogs in feeding trials, only cadmium was present in earthworms at concentrations anticipated as potentially adversely affecting frogs. Uptake of cadmium into earthworm tissue and consumption of soil retained in the gut yielded cadmium concentration in the frog diet that exceeded soil cadmium concentrations. Other metals such as copper and lead were not accumulated into earthworm tissue to the same extent as cadmium, but all metal residues increased in tissues as soil concentrations increased. Zinc was characterized at high soil concentrations by relatively low tissue concentrations, while at low soil concentrations zinc occurred in frog tissues at relatively high concentrations. This finding for zinc in worm tissue is probably due to that metal being an essential element; copper displayed a similar trend. Chromium concentrations in earthworms remained relatively low but increased in relative proportion to soil levels.

TABLE 1 -- *Metal concentrations in soils (μg/g; dry weight basis; mean ± standard deviation; n = 5).*

Treatment*	Zn	Cu	Cd	Cr	Pb
100 %	720 (114)	55 (7)	125 (20)	180 (38)	35 (2)
50 %	402 (12)	32 (5)	74 (3)	94 (1)	32 (2)
25 %	187 (3)	16 (4)	32 (1)	43 (1)	30 (2)
12.5 %	95 (4)	13 (2)	15 (1)	24 (1)	29 (1)
Reference soil	21 (1)	12 (5)	1 (0.1)	8 (1)	28 (2)
Culture soil	6 (1)	6 (3)	0.4 (0.4)	6 (6)	15 (2)

* % Site-soil included in exposure mix (site-soil amended with reference soil to yield 12.5%; 25%; and 50% mixtures).

TABLE 2 -- *Metal concentrations in earthworms (μg/g; dry weight basis; mean ± standard deviation; n = 10).*

Treatment*	Zn	Cu	Cd	Cr	Pb
100 %	508 (109)	33 (4)	609 (82)	83 (27)	25 (4)
50 %	310 (61)	22 (4)	246 (59)	46 (13)	21 (7)
25 %	203 (36)	16 (3)	72 (35)	23 (8)	18 (6)
12.5 %	148 (12)	14 (2)	37 (17)	14 (4)	18 (4)
Reference soil	134 (51)	14 (3)	2 (0.7)	6 (2)	15 (7)
Culture soil	104 (7)	14 (1)	2 (0.5)	5 (1)	14 (5)

* % Site-soil included in exposure mix (site-soil amended with reference soil to yield 12.5%; 25%; and 50% mixtures).

Metal Uptake in Amphibians

Amphibians exposed for 28 days to different metal-contaminated earthworm diets displayed no mortality nor exposure-related weight changes (data not shown). Frogs were exposed for 28 days to a maximum of approximately 600 μg cadmium/g in earthworms that had been previously exposed to soil contaminated with 125 ppm cadmium. In frogs, earthworm diets characterized by high cadmium concentrations resulted in exposure-related increases in cadmium in frog liver and egg samples. Relatively little cadmium, however, was present in thigh muscle. Residues for other metals in frog tissues were relatively low (Table 3).

Table 4 presents metal concentrations in frog liver. Zinc and copper were found at higher concentrations than other metals, probably due to their nutritional role in the frog. Chromium and lead were found only at low levels in liver tissues across all treatment groups, and concentrations were not highly correlated with exposures to those metals. At higher soil concentrations, cadmium was the only metal significantly elevated in liver, and the increase suggests the metal is accumulating in liver tissue during the 28-day feeding trial.

Table 5 summarizes results of metals analyses on egg samples from gravid females from 100% soil and 25% soil exposure groups. Cadmium appeared to be significantly elevated in eggs from the 100% exposure versus the 25% exposure, although these results must be considered preliminary owing to small sample size. Nonetheless, cadmium residues in egg samples appear increased.

Table 6 summarizes daily dose estimates for metals in the frog diet. Consumption of earthworms by frogs was similar across all treatments, in part owing to the relatively high variability in consumption rates among frogs in cadmium treatments. As noted earlier, frog body weights were similar among treatments (40.5 ± 6.4 gram; mean $\pm$ standard deviation), and no mortality was observed during the 28-day dietary exposure. Although the work summarized here was not designed as a reproduction study, future work should be completed to address potential reproductive effects associated with dietary exposures to cadmium. Cadmium dose is clearly dependent on soil cadmium concentration and cadmium residues in earthworm tissue, although not in a strictly linear manner. For example, cadmium consumption in the frog diet and cadmium dose are highly correlated (Spearman Rank Correlation, $r_\rho = 0.829$; $P = 0.042$), which suggests uptake efficiencies for cadmium across the gut of the frog are relatively high in these exposures. Overall, liver and egg appear to be the primary sinks of metal accumulation in this study.

Table 7 summarizes the earthworm-soil ratios derived from these studies. While only a first estimate of the relationships between metals in soil and metal residues in earthworms, these ratio estimators are equivalent to screening level ATFs in food-chain analyses, and may be applicable to the derivation of daily dose estimates or metal body burdens when tissue residues are absent from a data set and only soil concentrations are available.

TABLE 3 -- *Metal concentrations in amphibians: thigh muscle (µg/g; dry weight basis; mean ± standard deviation, n = 10)*

Treatment*	Zn	Cu	Cd	Cr	Pb
100 %	110 (28)	6 (1)	nd ...	1 (0.2)	nd ...
50 %	96 (14)	8 (0.8)	0.3 (0.1)	1.7 (0.6)	4.2 (0.5)
25 %	101 (23)	7 (0.9)	0.3 (0.1)	1.8 (0.4)	3.3 (0.8)
12.5 %	100 (28)	6 (1.2)	nd ...	1.9 (0.6)	3.2 (0.8)
Reference soil	100 (22)	6 (0.8)	nd ...	2.0 (0.3)	2.9 (0.8)
Culture soil	100 (19)	8 (1.5)	0.6 (0.4)	2.2 (0.3)	nd ...

* % Site-soil included in exposure mix (site-soil amended with reference soil to yield 12.5%; 25%; and 50% mixtures); nd = none detected.

TABLE 4 -- *Metal concentrations in amphibians: liver tissue (µg/g; dry weight basis; mean ± standard deviation, n = 10).*

Treatment*	Zn	Cu	Cd	Cr	Pb
100 %	114 (31)	196 (118)	8 (5)	nd ...	nd ...
50 %	94 (21)	241 (172)	3.5 (1.3)	2.0 (0.9)	7.3 (3.2)
25 %	109 (20)	263 (232)	1.9 (0.6)	2.6 (1.3)	8.1 (3.3)
12.5 %	94 (19)	369 (228)	2.2 (0.9)	2.8 (1.0)	7.7 (2.8)
Reference soil	96 (27)	245 (185)	1.1 (0.6)	2.7 (0.8)	6.5 (1.9)
Culture soil	106 (46)	213 (130)	1.0 (0.3)	3.1 (1.0)	4.6 (0.8)

* % Site-soil included in exposure mix (site-soil amended with reference soil to yield 12.5%; 25%; and 50% mixtures); nd = none detected.

TABLE 5 -- *Metal concentrations in amphibians: eggs (μg/g; dry weight basis).*

Treatment	Zn	Cu	Cd	Cr	Pb
100 %	248 (39)	7.0 (0.8)	6.4 (4.1)	0.8 (0.2)	nd ...
50 %	...[†]	...	...	...	...
25 %	215 (45)	7.2 (1.6)	0.5 (0.5)	1.2 (0.1)	1.4 (0.8)
12.5 %	...[†]	...	...	...	...

* % Site-soil included in exposure mix (site-soil amended with reference soil to yield 12.5%; 25%; and 50% mixtures); nd = none detected.
[†]No eggs were present in females for collection and analysis.

TABLE 6 -- *Earthworm consumption by frogs and cadmium accumulation by frogs in 'Soil - earthworm - amphibian' feeding study.*

Treatment*	Total earthworm consumed by frogs (total g wet weight[†])	Total Cd consumed (total μg[‡])	Concentration of cadmium in samples		
			Liver Cd (μg/g[‡])	Thigh Cd (μg/g[‡])	Egg Cd (μg/g[‡])
100 %	8.7 (3.9)	826 (378)	8.0 (4.4)	nd	6.4 (4.1)
50 %	9.9 (1.0)	378 (37)	3.5 (1.3)	0.3 (0.1)	no eggs
25 %	8.7 (2.0)	283 (65)	1.9 (0.6)	0.3 (0.1)	0.5 (0.5)
12.5 %	7.7 (3.7)	193 (91)	2.2 (0.9)	nd	no eggs
Reference soil	9.6 (0.8)	198 (16)	1.1 (0.6)	nd	no eggs
Culture soil	6.9 (0.4)	142 (7)	1.0 (0.3)	0.6 (0.4)	no eggs

[†] dry weight/wet weight ratio, 0.156; 0.158; 0.161; 0.167; 0.158; and 0.198 for 100%; 50%; 25%; 12.5%; reference, and culture soils, respectively.
[‡] dry weight basis.
* % Site-soil included in exposure mix (site-soil amended with reference soil to yield 12.5%; 25%; and 50% mixtures).

TABLE 7 -- *Earthworm - soil ratios or ATFs for Zn, Cu, Cd, Cr and Pb in amphibian feeding study.*[†]

Treatment*	Zn	Cu	Cd	Cr	Pb
100 %	0.71	0.60	4.9	0.46	0.71
50 %	0.77	0.69	3.3	0.49	0.66
25 %	1.08	1.00	2.3	0.32	0.60
12.5 %	1.56	1.08	2.4	0.58	0.62
Reference soil	6.38	1.17	2.0	0.75	0.54
Culture soil	17.3	2.30	5.0	0.83	0.93

[†] calculated as ratio of "concentration in earthworm/concentration in soil" from data in Tables 1 and 2.
* % Site-soil included in exposure mix (site-soil amended with reference soil to yield 12.5%; 25%; and 50% mixtures).

Discussion

Overall, the findings reported here can be focused on the transfer of metals, and in particular cadmium, from a loam or sandy loam soil to amphibians through a linear food-chain; that is, soil → soil invertebrate → amphibian. Given the focus on cadmium then,

- earthworm-soil ratios or ATFs for Cd varied from 2.3 to 4.9 across treatments of 12.5%, 25%, 50%, and 100% site-soil, but was relatively constant within treatments
- amphibian-earthworm ratios based on muscle residues for Cd varied from "not calculable" to 0.004 across treatments but was relatively constant within treatments
- Liver Cd concentrations >> Tissue (muscle) Cd concentrations
- Eggs appeared to be a relatively significant "biological sink" for Cd in females, but the effects of Cd on ovogenesis was not characterized.

Based on the data collected in this study, uptake of chromium, lead, and zinc by soil invertebrates appears to be relatively low in the loam or sandy loam soils used in these studies. Estimates of ATF for these metals were less than that for cadmium, although at low concentrations ATFs for the nutrients copper and zinc were greater than 1. Cadmium accumulation, however, appears to be of relatively greater significance from an ecotoxicological perspective; its presence in the diet as either tissue residues or as residual soil in the gut contribute to the cadmium dose in frogs consuming earthworms exposed to cadmium-contaminated soil. From a food web perspective, cadmium in exposed earthworms was associated with a soil ATF much greater than 1.0 which suggests it would probably be readily bioavailable to predators of soil invertebrates. In contrast, lead levels in soil were relatively low (0.6 to 0.7) for an ATF for soil → soil invertebrate transfers in a loam to sandy loam soil; that is, lead levels in earthworms did not exceed soil levels in any treatment group.

The present study's results are consistent with those reported by Ma (1982) where cadmium was taken up by earthworms to a higher concentration relative to substrate concentration than other heavy metals. Likewise, Woodyard and Haufler (1991) observed that heavy metal tissue residues in birds fed earthworms varied directly with the concentration of metal in contaminated feed. In their study, metal residues for cadmium, chromium, copper, and zinc were all significantly higher in worms from soil bearing higher metal concentrations, but only cadmium was accumulated to concentrations greater than those observed in soil. That is, cadmium ATFs derived from the work of Woodyard and Haufler (1991) were also greater than 1.

Although bioaccumulation is conceptually similar in both terrestrial and aquatic habitats, the role of bioaccumulation and food-chain analysis in ecological risk assessment differs across these habitats. Matrix differences directly or indirectly influence the extent and rate of bioaccumulation in various habitat settings. For example, physiological differences between receptors (aquatic versus terrestrial) influence bioaccumulation. Consequently, the data required to complete food-chain analysis are not always available for use in predictive models when aquatic and terrestrial habitats are being considered.

In terrestrial habitats, bioaccumulation is generally dominated by dietary exposures unless contaminants readily volatilize from a solid or liquid phase into air or are readily absorbed across dermal epithelia, which may be critical in evaluating ecological risks within the rhizosphere or for soil macroinvertebrates. Many variables affect the magnitude of bioaccumulation in terrestrial receptors, and the transfer of chemicals within food chains may conveniently be described by transfer coefficients or functions such as ATFs that characterize the relationships between trophic levels in food chains or in food webs. For example, such factors may characterize the transfer of chemicals from soil to water or from soil and sediment to an organism dwelling within these matrices up the food chain. For the most part, these variables are, at best, empirical descriptors that characterize the transfer of chemicals from an environmental matrix (soil, sediment, or water) to microorganisms, plants, or animals. In turn, the microorganisms are characterized by the transfer coefficients or functions that describe the transfer of chemicals from these biological matrices to the consumers positioned at higher trophic levels. Alternatively, modeled transfer coefficients may be used to describe these transitions, but the variance component is often poorly described or missing.

In soils and sediments, the initial transfer functions are influenced by physicochemical characteristics of the matrix such as particle size distribution or texture, cation exchange capacity, degree of carbon enrichment, and porosity. As previously observed (Pascoe et al. 1994, Pascoe et al. 1996, Linder et al. 1992), the potential for overestimating or underestimating these transfer functions increases as the complexity of the environmental mixture increases. For example, the transfer of chemicals within tissues is often enhanced by the lipid content of a tissue or by physiological functions that may predispose an organism to accumulate environmental chemicals (Hamelink et al. 1996).

Biological processes such as depuration (clearance), detoxification, and other biotransformation mechanisms affect the potential for bioaccumulation within an organism; however, these physiological and biochemical functions are infrequently incorporated into a food-chain analysis unless uncertainty is not an issue.

Conclusions

These results clearly indicate that the transfer of chemicals among trophic levels is critical to the characterization of the bioaccumulation process in wetland and terrestrial food-chains. The process is strongly influenced by numerous interacting abiotic and biotic factors, including the physicochemical properties of soil, and the role, if any, that the metal has in the receptor as a required trace element.

Food-chain analysis is only one of the tools available to evaluate ecological risks; assessment strategies have been developed that consider these vertebrates under various regulatory drivers (Linder et al. 1993, Pascoe et al. 1994a,b). Wildlife may be exposed to a wide variety of anthropogenic chemicals that are accidentally or intentionally released into their environment. While these ecological receptors may not always be the most sensitive targets during a risk assessment exposure evaluation, terrestrial and semiaquatic vertebrates are critical components within habitats at risk and must be considered as part of the process. Food-chain modeling is only one way of integrating ecological information into the risk assessment process, especially for contaminants that present a hazard associated with long-term, less-than-acutely toxic exposures (for example, chemicals that bioconcentrate or bioaccumulate). If used with caution, these conceptually simple models may help focus the analysis of risks on ecological targets or receptors potentially exposed to chemicals in various environmental media (for example, soils and sediment) or food sources.

Acknowledgments

The authors would like to thank Keith Hoddinott for organizing each of the three symposia in the series (*Superfund Risk Assessment in Soil Contamination Studies*). The senior author also extends his thanks to James Fairchild with the US Geological Survey, Biological Resources Division, Environmental and Contaminants Research Center, Columbia, MO, for travel support to attend the third symposium. Thanks also go to Judith Bauer, who reviewed preliminary drafts of this manuscript, and to Margaret Linder for continuing insights into the role of biological mechanisms that contribute to the analysis of ecological risks in wildlife.

References

Alloway, B.J., Ed., 1990, *Heavy Metals in Soils*, Blackie/John Wiley & Sons, Inc., New York, NY.

American Society for Testing and Materials (ASTM). 1998. "Standard Guide for Conducting Laboratory Soil Toxicity or Bioaccumulation Tests with the Lumbricid

Earthworm *Eisenia fetida* (E-1676)," American Society for Testing and Materials, ASTM Committee E-47 on Biological Effects and Environmental Fate, West Conshohocken, PA.

EPA, 1989, *Protocols for Short-Term Toxicity Screening of Hazardous Wastes,* U.S. Environmental Protection Agency, Corvallis Environmental Research Laboratory, Corvallis, OR. 600/3-88/029.

Hamelink, J.L, Landrum, P.F., Bergman, H.L. and Benson, W.H., Eds., 1994, *Bioavailability: Physical, Chemical, and Biological Interactions.* Lewis Publishers, Inc., Boca Raton, FL.

Ireland, M.P., 1977, "Lead Retention in Toads *Xenopus laevis* Fed Increasing Levels of Lead-Contaminated Earthworms," *Environmental Pollution* Vol. 12, pp.85-92.

Klute, A., Ed. 1986, *Methods of Soil Analysis. Part 1 - Physical and Mineralogical Methods.* 2nd ed., American Society of Agronomy and Soil Science of America.

Linder, G., Callahan, C. and Pascoe, G., 1993, "A Strategy for Ecological Risk Assessments for Superfund: Biological methods for evaluating soil contamination," In *Superfund Risk Assessment in Soil Contamination Studies, ASTM STP 1158,* Keith B. Hoddinott, Ed., American Society for Testing and Materials, Philadelphia, PA., pp. 288-308.

Linder, G., Hazelwood, R., Palawski, D., Bollman, M., Wilborn, D., Malloy, J., DuBois, K., Ott, S., Pascoe, G. and DalSoglio, J., 1994, "Ecological Assessment for the Wetlands at Milltown Reservoir, Missoula, Montana: Characterization of emergent and upland habitats," *Environmental Toxicology and Chemistry,* Vol. 13, pp. 1957-1970.

Lyon, J.G., 1993, *Practical Handbook of Wetland Identification and Delineation,* Lewis Publishers, Boca Raton FL.

Ma, W.C., 1982, "The Influence of Soil Properties and Worm-Related Factors on the Concentration of Heavy Metals in Earthworms," *Pedobiologia,* Vol. 24, pp. 109-119.

Pascoe, G.A., Blanchet, R.J., Linder, G., Palawski, D., Brumbaugh, W.G., Canfield, T.J., Kimble, N.E., Ingersoll, C.G., Farag, A. and DalSoglio, J.A., 1994a, "Characterization of Ecological Risks at the Milltown Reservoir-Clark Fork River Sediments Superfund Site, Montana," *Environmental Toxicology and Chemistry* Vol. 13, pp. 2043-2058.

Pascoe, G.A., Blanchet, R.J. and Linder, G., 1994b, "Bioavailability of Metals and Arsenic to Small Mammals at a Mining Waste-Contaminated Wetland," *Archives of Environmental Contamination and Toxicology*, Vol. 27, pp. 44-50.

Pascoe, G.A., Blanchet, R.J., and Linder, G., 1996, "Food Chain Analysis of Exposures and Risks to Wildlife at a Metals-Contaminated Wetland," *Archives of Environmental Contamination and Toxicology*, Vol. 30, pp. 306-318.

Woodyard, D.K. and Haufler, J.B., 1991, "Transfer of Sludge-Borne Metals Through a Woodcock-Earthworm Food Chain," In *Risk Evaluation for Sludge-Borne Elements to Wildlife*, Garland Publishing, Inc., New York, NY.

Risk Management

David E. Mohr[1] and Fredrick S. Illich[1]

PROCEDURAL AND COMPUTERIZED METHODS FOR ASSESSING
SITE RISKS TO ACHIEVE REMEDIATION GOALS

REFERENCE: Mohr, D.E. and Illich, F.S., "Procedural and Computerized
Methods for Assessing Site Risks and Achieving Remediation Goals," *Superfund
Risk Assessment in Soil Contamination Studies: Third Volume, ASTM STP 1338,*
K.B. Hoddinott, Ed., American Society for Testing and Materials, 1998.

ABSTRACT: A methodology was developed to characterize risks at hazardous
waste sites to expedite evaluation of remedial alternatives in various types of
environmental media. This methodology is composed of standard operating
procedures that define the data collection and evaluation process and is assisted
by a computerized system designed to maximize the analysis of large quantities of
environmental data. The procedural process incorporates regulatory guidance and
study plan design. The computerized data management system incorporates data
qualification, data reduction, and risk analysis. Together, the process and the
supporting system provide a framework used to rapidly process and analyze
environmental data through risk characterization. The results from risk
characterization are then directly used to formulate and evaluate remedial
alternatives. This approach significantly streamlines the work necessary to
establish remediation goals in compliance with environmental regulation.

KEYWORDS: Hazardous waste, data management, data quality assessment, risk-
based screening, chemicals of potential concern, remedial action criteria

---

The life cycle of environmental data encompasses an extensive range
beginning with planning for the collection of data, continuing through the
collection and evaluation of data, and ending in the development of remedial
alternatives based on risk assessment results. For instance, at the planning stage
for data collection, analytical methods are selected to perform chemical analysis
of hazardous substances based on risk-based screening concentrations derived for
each chemical and environmental medium to be evaluated. Data generated from
laboratory analysis of samples collected from the various environmental media are
loaded into an environmental database for processing. Automated data

---

[1]Environmental Engineer, Data Management and Analysis Department, URS
Greiner, 2401 Fourth Avenue, Suite 1000, Seattle, WA 98121.

qualification and reduction are then performed in accordance with the objectives established in the data collection plans and regulatory guidelines for data quality assessment for risk assessments. From this, chemicals of potential concern are identified for further study based on frequency and concentration of measured values. These chemicals relate to sites and environmental media that may require remediation based on calculated risks.

Planning for Data Collection

During the planning phase of a project, the basis for an environmental study is formulated in a sampling and analysis plan. Components of this plan are based on the selection of ASTM standards used in environmental studies in accordance with ASTM Site Characteristics for Environmental Purposes with Emphasis on Soil, Rock, the Vadose Zone, and Groundwater (D5730-95a). Data related issues at this phase involve development of data quality objectives, formulation of sampling rationale, delineation of sampling locations, enumeration of sample quantities and types, determination of quality control samples, specification of laboratory analysis requirements, and identification of sample inventory control to be used consistently throughout the project (U.S. EPA 1988).

Data quality objectives establish the goals for measurement thresholds, precision, accuracy, completeness, comparability, and representativeness of the data to be collected on the project (U.S. EPA 1993b, 1987). Since some regulations and risk-based screening levels are set at low concentrations for toxic contaminants, an evaluation of analytical method detection limits is necessary. To assist the evaluation, computerized chemical libraries are accessed to match risk-based goals and data quality objectives with specific performance parameters of analytical methods.

An evaluation of chemical screening criterion values and analytical method reporting limits (U.S. EPA 1994a) for soil is conducted (Table 1). This evaluation establishes reporting limit goals and determines the ability of the analytical methods to achieve these goals for individual chemicals of interest. Chemical screening criterion values are first derived for individual chemicals for a particular environmental medium. Derivations are based on an evaluation of the chemical concentrations for carcinogenic and noncarcinogenic risk-based screening concentrations (RBSCs), site-specific ecological RBSCs, regulatory chemical concentration thresholds, and regional-specific background concentrations measured for inorganics. The chemical screening criterion values derived for soil are then set as the reporting limit goals for the analytical methods.

Both the screening criterion value and the corresponding screening criterion source are shown for each chemical (analyte) in which values are derived. The chemicals are then grouped by analytical method, within which each analytical method reporting limit is shown for each chemical as published in the referenced analytical method. The screening criterion values are then compared to the method reporting limits. If the method reporting limit is greater, the reporting limit goal is not achieved (decision column contains "No"). If the

TABLE 1—*Evaluation of Analytical Method Reporting Limit Goals*
Matrix: Soil   Units: mg/kg

Analytical Method Analyte Name	Method Reporting Limit	Screening Criterion Value	Screening Criterion Source	Reporting Limit Goal Achieved?	Exceedance Ratio
Inorganics					
Antimony	12	11	Human Health RBSC	No	1.1
Arsenic	2	7.5	Background	Yes	...
Barium	40	1920	Human Health RBSC	Yes	...
Beryllium	1	0.67	Background	No	1.5
Cadmium	1	13.4	Background	Yes	...
Chromium	2	625	Ecological RBSC	Yes	...
Copper	5	98	Background	Yes	...
Lead	1	34	Ecological RBSC	Yes	...
Mercury	0.2	1.6	Background	Yes	...
Silver	2	137	Human Health RBSC	Yes	...
Zinc	4	80.3	Background	Yes	...
Nitroaromatics and Nitramines					
1,3,5-Trinitrobenzene	0.25	1.37	Human Health RBSC	Yes	...
1,3-Dinitrobenzene	0.25	2.74	Human Health RBSC	Yes	...
2,4,6-Trinitrotoluene	0.25	2.13	Human Health RBSC	Yes	...
2,4-Dinitrotoluene	0.25	0.0939	Human Health RBSC	No	2.7
2,6-Dinitrotoluene	0.26	0.0939	Human Health RBSC	No	2.8
Nitrobenzene	0.26	3.6	Ecological RBSC	Yes	...
RDX	1	0.23	Ecological RBSC	No	4.3

method reporting limit is less than or equal to the screening criterion value, the reporting limit goal is achieved (decision column contains "Yes").

In cases where the reporting limit goal is not achieved, an exceedance ratio is calculated. This ratio indicates the magnitude by which the method reporting limit exceeds the screening criterion value (reporting limit goal). The magnitude of the exceedance ratio is further evaluated in relation to the method reporting limits. This evaluation considers laboratory techniques that may be used to achieve lower reporting limits. Even though the reporting limits most probably will not be met for some chemicals, the specifications for lower reporting limits are applied to approach the reporting limit goals as much as possible. The benefit of this approach is the reduction of the uncertainty of whether the chemical is present in the environment at a concentration that could cause risks to receptor populations.

Data Collection and Validation

Following the planning process, field data and environmental sample collection are performed in accordance with standard operating procedures (URS 1995a, 1995b). These procedures include step-by-step instructions for various types of sample collection and corresponding documentation. Field data are transcribed onto predesigned data collection forms that contain definitions of all sample inventory control information to be used on a project in accordance with specifications in the sampling and analysis plan.

Samples are transmitted to analytical laboratories for various chemical analysis (U.S. EPA 1994a) and physical analysis (ASTM D5730-95a). This stage of data collection focuses on the correct specification of analytical methods for sample analysis, the qualification of data generated by the laboratory in accordance with data quality objectives, and the appropriate application of the chemical data results to the environmental study. All data components, including laboratory methods, environmental matrices, sample numbers, etc., are used consistently providing common terminology among the field teams, project management, subcontract procurement, laboratories, and data validation companies. Procedures are followed to evaluate and verify whether the work performed by the analytical laboratories meet specifications in the data collection plan. Data are independently validated and qualifiers are assigned to data consistent with the data validation functional guidelines (U.S. EPA. 1994b, 1993a).

Environmental Data Management System

In order to evaluate the chemical concentrations obtained from environmental sample analysis, a process was developed by which geographical, geological, and chemical data are organized, managed, analyzed, and presented for hazardous waste site characterization. The goal was established to facilitate rapid and accurate recording of field observations and the uniform collection and analysis of complete and traceable data that ultimately depict the environmental condition of a site. This process comprises standardized methods for measuring field parameters, collecting samples from a variety of environmental matrices, and using an integrated set of data collection forms and formats to record information in accordance with data collection plans.

For an average-sized hazardous waste study, thousands of discrete measurements are obtained from field surveying and by laboratory sample analysis. To accommodate this large database, an automated system is used to manage the resulting data. At the core of this system is a simplistic computer file structure that is a direct image of the data collection process. Simplicity of the system permits the creation of a portfolio of data analysis and presentation utilities (Figure 1). These utilities are connected to the system radially as processing (application) modules that operate from the core system as follows:

- Field sampling and analysis plan modules.
- Analytical method library containing all chemicals and properties.
- Regulatory and risk-based criteria for comparison to measured data.
- Sample tracking from field collection through laboratory analysis.
- Data entry using computer screens modeled from field forms.
- Data import utilities from laboratory sample analysis.
- Site maps indicating sampling locations and chemical concentrations.
- Report generation from menu selection of preprogrammed tables.
- Data transfer utilities to other computer databases and models.
- Document control for archiving field and laboratory data.

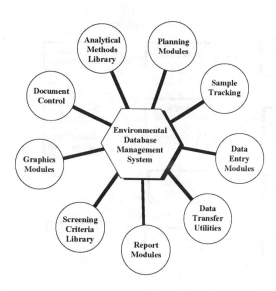

FIG. 1—*Environmental Database Management System*

Environmental data from hazardous waste projects are not only great in number, but also diverse in type. These data frequently cover a large geographical area, a variety of environmental media, a variable quantity of physical and chemical measurements, and a variable span of time. As a result, the system architecture and interconnectivity of the supporting utilities is designed to specifically address these conditions, yet remain flexible to accommodate diversity and change in data evaluation techniques typically driven by regulation concerning environmental protectiveness.

The successful implementation of the data management process is founded in standard operating procedures for data collection and data management. These procedures establish uniform methods for the collection and recording of data. Data management is then the process which maintains and coordinates the organization of the data. Planning for the collection and generation of data (sampling and analysis plans) also uses the procedures as a basis for uniformly itemizing the samples to be collected. Overall, this process ensures the integrity and rapid processing of data. The data flow process illustrates the organization of data (Figure 2).

Data Quality Assessment

Following chemical analysis and assembly of data in the computerized data management system, an assessment of data quality is conducted prior to data analysis and risk evaluation. A method was developed for evaluating the overall data quality from environmental studies in terms of established data quality

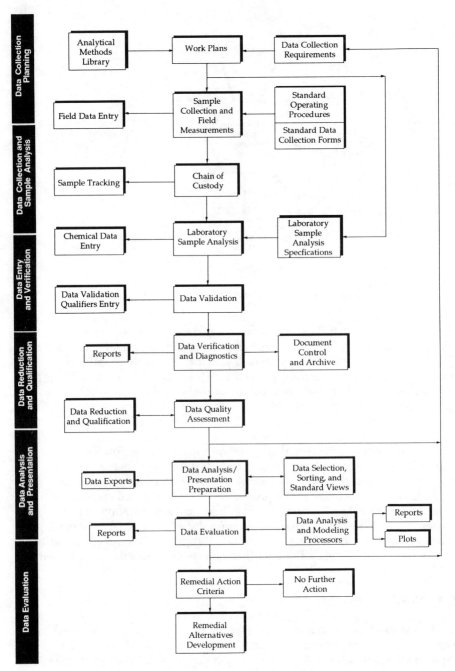

FIG. 2—*Environmental Data Flow Diagram*

objectives and the effectiveness of the data collection and sample analysis processes. This method includes the components for assessing the quality of environmental data used in risk assessments (U.S. EPA 1990). The assessment of data quality spans all elements defined in quality assurance project plans and all principal analytical methods that are typically employed for sample analysis of chemical concentrations in various environmental media. The performance of the overall data collection process is evaluated in terms of the completeness of the project plans, effectiveness of field measurement and data collection procedures, and relevance of laboratory analytical methods used to generate data as planned. The goal of the data quality assessment is to establish data integrity and data usability for the intended purpose of study.

Data quality assessments are prepared to document the overall quality of data collected in terms of the established data quality objectives and the effectiveness of the data collection and generation processes. The quality of data measured is assessed to ensure that data are scientifically valid, of known and documented quality, and technically defensible. The result of the data quality assessment is to achieve an acceptable level of confidence in the decisions that are to be made from measurements and data by controlling the degree of total error permitted in the data through quality control checks. Data that fail the quality control checks or do not fall within the acceptance criteria established are qualified for limited use or rejected from further use.

The six steps in the data quality assessment progressively reduce the collected environmental data to a set of valid data of known and documented quality. Each step relies on a combination of computerized analysis and technical evaluation using rules defined by data validation and goals established in the sampling and analysis plan to assess data quality. For example, a summary of data qualifiers assigned to each individual chemical result for each analytical method is prepared (Table 2). This summary shows the quantities and percentages of data qualifiers and is used to determine whether a sufficient population of data exists for environmental study based on established quality goals. The resulting set of data is then available to use with confidence in environmental characterizations, risk assessment, and remedial decisionmaking.

Data Qualification and Reduction Procedures

A component of the data quality assessment is automated data qualification and reduction in accordance with a set of procedures. Chemical data are reduced to the set of values within data management system that meet minimum quality criteria and can be considered the "best" value within a set of data. Data reduction is performed as either the process of elimination by which data not meeting the minimum criteria are not used, or a process of selection by which data are designated as the best value for a set of two or more values. Multiple values typically result from laboratory duplicates or dilutions that are generated during the normal process of laboratory sample analysis.

Following data validation, the data set is further reviewed for appropriate application of data qualifiers. The purpose of this review is to apply consistent

TABLE 2—*Summary of Data Validation Qualifiers*
Analytical Method: Inorganics

Analyte Name	A	A%	J	J%	R	R%	U	U%	UJ	UJ%	Total
Antimony	0	0	2	4	0	0	11	24	32	71	45
Arsenic	42	91	3	7	0	0	1	2	0	0	46
Barium	44	98	0	0	0	0	1	2	0	0	45
Beryllium	43	96	0	0	0	0	2	4	0	0	45
Cadmium	16	36	0	0	0	0	29	64	0	0	45
Chromium	45	100	0	0	0	0	0	0	0	0	45
Copper	45	100	0	0	0	0	0	0	0	0	45
Lead	26	57	20	43	0	0	0	0	0	0	46
Mercury	16	35	0	0	0	0	30	65	0	0	46
Silver	1	2	0	0	0	0	44	98	0	0	45
Zinc	45	100	0	0	0	0	0	0	0	0	45

rules for qualification of data independent of the laboratories and data validators. The data qualification review specifically involves examining the full data set to determine which data represents the environmental condition for the intended purpose of study. Automated data qualification and reduction is used to select the best value for each chemical or group of chemicals by the following methods:

- Resolve multiple valid chemical values caused by laboratory dilutions, reanalyses, and duplicates.
- Resolve multiple valid values caused by chemicals measured using multiple analytical methods for the same sample.
- Resolve multiple valid values caused by collection and analysis of field duplicates.
- Compute total values for classes of compounds.
- Normalize sediment data based on organic carbon content.

Risk-Based Screening and Statistical Analysis

Following data diagnostics and reduction, the process of data analysis begins. Patterns of analysis have evolved into standard computerized reports and menus used to extract and array data in ways that support an environmental study or remediation design. Data analysis is an iterative process whereby the data are examined at different levels in the data management system and compared to various criteria to ascertain the significance of the environmental measurements. Typical reports include presentation of the quantities of samples collected for different environmental media or methods, statistical summaries by analytical method and chemical, and comparison of data to regulatory or risk-based screening levels. Following routine comparisons, more in-depth views of the data are performed, examining the distribution of a specific chemical for a specific media and geologic strata.

Environmental projects typically have a number of criteria for which the data are compared. These criteria are applied to detected analytes for the purpose of identifying chemicals of potential concern (COPCs). For instance, some COPC screening processes list only data that exceeds a natural background level and risk-based screening criteria for a particular chemical and environmental matrix. Since the selection of the screening criteria is usually performed in multiple steps, comparison of environmental data to regulatory and risk-based criteria can be complex to perform.

Essentially, two criteria are used in the risk assessment for identifying COPCs in each environmental medium. For inorganics, the maximum concentration for a given chemical must exceed both the statistical background concentration and the calculated RBSC. For organics, the maximum concentration for a given chemical must exceed the calculated RBSC. Detected chemicals are treated at the measured value while non-detected chemicals are treated at half of the detection limit to reduce the potential for underestimating risks (U.S. EPA 1991b, 1989).

To simplify this task, a screening criteria library was developed containing a collection of chemicals and values which are commonly compared to environmental site data. Types of criteria contained in the library include background levels, practical quantitation limits, regulatory action levels, technically achievable levels, and RBSCs. These values, either derived during a risk assessment or published by regulatory agencies, are associated by environmental matrices and units of measure. Each set of screening criteria is designated by an abbreviation, a screening criteria type, description of how comparisons to measured values should be made, and a list of chemicals and concentrations applicable for the criteria. The chemicals are designated by the chemical name and a unique identifier. This identifier is the key identifier for comparing a set of screening criteria to a set of measured chemical values within the environmental data management system.

Once COPCs are selected, reasonable maximum exposure concentrations are computed as an input into the models used for determining risk (Table 3). The derivation of the reasonable maximum exposure is based on choosing the maximum detected value or the 95 percent upper confidence limit of the mean, whichever is lower. The risk evaluation statistics are generated by the data management system using the measured chemical concentrations.

Site Screening Based on Risk Assessment

Following the selection of COPCs and the determination of risks in the various environmental media under study, sites are subjected to a screening process. The screening is designed to identify sites and media that may require remediation based on calculated risks. Sites are subjected to a two-phased screening process to identify those that require detailed evaluation of remedial alternatives. Phase 1 screening involves evaluating sites in relation to calculated risks while Phase 2 screening considers site characteristics and environmental setting that warrant and could support remedial evaluation.

TABLE 3—*Detected Analytes and Risk Evaluation Statistics*
Matrix: Soil   Units: mg/kg

Analyte Name	Quantity Tested	Quantity Detected	Minimum Detected Value	Maximum Detected Value	Average Exposure Concentration	95% Upper Confidence Limit	Reasonable Maximum Exposure
Total Inorganics							
Antimony	22	2	9.7	14.7	4.62	5.58	5.58
Arsenic	22	22	0.74	5.1	2.22	2.65	2.65
Barium	22	22	12.7	116	32.8	43.5	43.5
Beryllium	22	22	0.55	1.3	0.857	0.928	0.928
Cadmium	22	4	0.86	2.5	0.694	0.888	0.888
Chromium	22	22	3.2	55.8	13.2	18.6	18.6
Copper	22	22	33.8	76.4	44.9	49.8	49.8
Lead	22	22	1.5	76.9	18.1	28	28
Mercury	22	4	0.07	0.38	0.67	0.097	0.097
Silver	22	14	4.4	19.2	6.93	8.89	8.89
Zinc	22	22	27.2	549	113	165	165

Phase 1 of screening utilizes the results of the human health and ecological risk assessments. Sites not exceeding acceptable risk threshold criteria are eliminated from further consideration since they do not pose significant adverse threat to human health and the environment. An evaluation for this phase of screening (Table 4) summarizes risk assessment results for each site and impacted environmental media within the site. The results summarized are the projected future land use for each site, the impacted environmental media evaluated in the risk assessment, the chemical classes of the COPCs driving the calculated risks, and the calculated risks for scenarios involving human and ecological receptors.

Human health and ecological risks are calculated in the risk assessment for each exposure scenario and pathway contributing risk. Human health risks are evaluated, where applicable, based on residential, industrial (occupational), and/or recreational land use. Risk thresholds are cumulative for a site receptor and do not include risk from background concentrations (i.e., incremental risks only) consistent with the risk assessment.

Remedial action is generally not warranted under regulatory guidelines where cumulative site human health carcinogenic risk , based on the reasonable maximum exposure for both current and future land use, is less than 10^{-4} and the noncarcinogenic risk is less than a hazard index of 1.0 (U.S. EPA 1991a). Calculation of ecological risk entails a high degree of uncertainty and use of conservative assumptions. As a result, a hazard index of 10 is considered to indicate a possibly adverse impact and selected as an acceptable risk threshold.

Phase 2 screening is used to evaluate sites where risks exceed the thresholds established in the first phase of screening. The screening criteria developed for this phase involves comparing site characteristics and environmental setting in relation to the propensity for calculated risks to realistically exist at a

TABLE 4—*Phase 1 Site Screening Based on Risk Assessment Results*

Site Designation and Site Type	Future Land Use	Impacted Environmental Medium	Chemical Classes of the Chemicals of Potential Concern	Human Health Risk Scenario						Ecological Risk Hazard Index	Retain Site for Phase 2 Screening?
				Residential		Recreational		Occupational			
				Cancer Risk	Hazard Index	Cancer Risk	Hazard Index	Cancer Risk	Hazard Index		
Site 4 Landfill	Recreational	Subsurface Soil	SVOC, P/A, TIN, D/F	3.6E-5	0.27	—	—	1.8E-6	0.04	—	Yes
		Surface Soil	P/A, TIN, D/F	—	—	—	—	—	—	120	
		Sediment	SVOC, P/A	1.1E-8	<0.01	1.1E-8	<0.01	—	—	20	
Site 10 Waste Storage	Industrial	Surface Soil	SVOC, P/A, TIN	3.4E-5	<0.001	5.1E-7	<0.001	1.7E-6	<0.001	29	Yes
Site 15 Waste Storage	Industrial	Sediment	SVOC, P/A, D/F	6.7E-8	<0.001	6.7E-8	<0.001	—	—	—	
		Soil	SVOC, P/A, D/F	4.2E-5	0.001	6.3E-7	<0.001	2.1E-6	<0.001	—	
Site 17 Fuel Storage	Industrial	Sediment	VOC, SVOC, P/A, TIN	1.1E-7	<0.001	1.1E-7	0.001	—	—	880	Yes
		Surface Soil	SVOC, P/A, TIN	1.6E-5	16	2.4E-7	0.007	8.0E-7	0.022	73	
		Surface Water	VOC, SVOC, P/A, TIN	1.7E-4	25	1.1E-6	14	—	—	1 370	
Site 23 Drum Storage	Recreational	Sediment	SVOC, TIN	3.3E-8	<0.01	3.3E-8	<0.01	—	—	49	Yes
		Surface Soil	P/A, TIN, SVOC	2.5E-6	6.4	3.7E-8	0.3	1.2E-7	0.3	—	
		Subsurface Soil	P/A, TIN	—	—	—	—	—	—	89	
Site 75 Product Storage	Industrial	Surface Soil	P/A, TIN	2.4E-6	<0.001	1.7E-8	<0.001	5.6E-8	<0.001	2.0	No
Site 76 Product Storage	Industrial	Surface Soil	SVOC, TIN, P/A	6.7E-5	0.70	7.5E-7	0.02	2.4E-6	0.06	13	Yes

Notes:

Chemical Classes:
VOC - volatile organic compounds   P/A - pesticides/Aroclors   TIN - total inorganics
SVOC - semivolatile organic compounds   D/F - dioxins/furans

site sufficient to pose adverse threat to human health and the environment. The factors considered during this phase are the following:

- Assumed exposure pathways
- Future land use
- Spatial distribution of chemical concentrations
- Magnitude and frequency of chemical concentrations
- Receptor populations
- Remedial actions already taken
- Potential habitat damage associated with remedial activities
- Magnitude of risks above the acceptable threshold
- Impact of factors causing overestimation or underestimation of risks
- Other site factors

Development of Remedial Action Criteria

Following site screening, remedial action criteria are developed for sites and site types. These criteria include the following elements (U.S. EPA 1988):

- Remedial action objectives that indicate where remedial actions may be needed and what they should accomplish.
- Preliminary remediation goals that provide targets to develop and evaluate remedial response alternatives in terms of specific chemical concentrations for individual environmental matrices.
- General response actions that are broad, generic categories of remedial actions appropriate for accomplishing remedial action objectives for a particular environmental medium, either by themselves, or in combination.

Remedial action criteria proposed for all affected sites are formulated for a particular environment medium and risk scenario to be protective of human health and the environment. Specifying the exposure route rather than just an acceptable concentration range is provided so that protectiveness can be achieved by preventing exposures (e.g., by containment or institutional controls) and by reducing concentrations of chemicals in the environmental media of concern. For each established remedial action objective, the following information is specified:

- Exposure routes and receptors of concern.
- Chemicals of concern.
- Acceptable concentration for each chemical exposure of concern.

Preliminary remedial goals are target chemical concentrations allowing development and evaluation of remedial response alternatives and are considered to be acceptable concentrations for each chemical to exist in a particular environmental medium. These goals are based on cancer risk for carcinogenic or a hazard index for noncarcinogens, and are expressed in chemical-specific terms such as a cleanup concentration. Acceptable concentrations are established by means of risk-based calculations and consideration of chemical-specific criteria and regulatory guidelines. Preliminary remedial goals provide targets for developing and evaluating remedial response alternatives but are considered preliminary until agreement regarding the remedial action is reached.

Sites retained from the screening process for development of remedial action criteria and detailed evaluation are listed (Table 5) along with the chemicals of concern and exposure pathways presenting excessive adverse risk. Chemicals of concern represent the risk drivers causing the exceedance of acceptable risk thresholds for a given site, impacted environmental medium, and receptor and are ranked in order of decreasing risk. These criteria are developed based on each applicable risk scenario for which impacted environmental media and chemicals of concern have been identified.

Summary

By adopting methodologies described in various regulatory agency guidance documents and developing standard operating procedures, an automated data management system was able to be developed for analyzing environmental data and assessing risks at hazardous waste sites. Both the procedures and computing environment used and operated in concert for processing large volumes of environmental data against risk-based criteria, facilitates the achievement of remediation goals—the logical outcome of waste site investigations. The benefits derived from this approach for characterizing risks and evaluating remedial alternatives are as follows:

- Data collection planning using comparisons made between computerized libraries of analytical method performance specifications and chemical screening criteria derived from regulatory and risk-based screening concentrations.
- Environmental data management conducted by an automated system designed to organize, analyze, and present geographical, geological, and chemical data within a single computing environment.
- Data quality assessment that incorporates data qualification and reduction procedures from which the data usability acceptance rate of over 99 percent has been sustained for 90 environmental studies.
- Risk-based screening and statistical analysis of measured chemical concentrations in the environmental data management system that rapidly identifies chemicals of potential concern for individual environmental media at specified sites.
- Site screening process designed to further evaluate sites that may require remediation based on calculated risks by facilitating the development of remedial action criteria.

TABLE 5—*Sites Retained for Development of Remedial Action Criteria*

Site Designation	Remedial Action Objectives	Impacted Environmental Media	Chemicals of Concern (Ranked in Descending Order of Risk)	Chemical Concentrations (mg/kg)	General Response Actions and Remedial Technologies
Site 4 Landfill	Environmental Protection Prevent ingestion of impacted subsurface soils by birds and uptake by invertebrates and plants	Subsurface soil	Zinc	67	No action
			Lead	34	Institutional controls: land use restrictions and periodic site monitoring
			2,3,7,8-TCDD	2×10^{-6}	
			Aroclor 1260	0.09	Containment: soil cover
			Aroclor 1254	0.09	
Site 17 Fuel Storage	Environmental Protection In the waste oil pond, prevent uptake of impacted freshwater sediments by aquatic organisms	Freshwater sediments	Fluorene	0.035	No action
			Aroclor 1260	0.005	Institutional controls: periodic site monitoring
			2-Methylnaphthalene	0.065	Containment: sediment cover
			Antimony	0.03	Source removal: excavation and disposal
			Phenanthrene	0.225	
			Lead	35	
			Acenaphthene	0.15	
			Pyrene	0.35	
			Mercury	0.15	
			Heptachlor	3×10^{-4}	
			Fluoranthene	0.6	
			Chrysene	0.4	
			Nickel	30	
			Zinc	120	
			Benzo(a)pyrene	0.4	
			Anthracene	0.085	
			Copper	70	

References

United States Environmental Protection Agency (U.S. EPA). 1987. *Data Quality Objectives for Remedial Response Activities: Development Process.* EPA-540/G-87/003.

————. 1988. *Guidance for Conducting Remedial Investigations and Feasibility Studies Under CERCLA.* OSWER Directive 9355.3-01. October 1988.

————. 1989. *Risk Assessment Guidance for Superfund.* Vol. 1, *Human Health Evaluation Manual (Part A).* Interim Final. EPA 540/1-89-002. Office of Emergency and Remedial Response. December 1989.

————. 1990. *Guidance for Data Useability in Risk Assessment.* EPA/540/G-901008.

————. 1991a. *EPA Region 10 Supplemental Risk Assessment Guidance for Superfund.* August 1991.

————. 1991b. *Role of the Baseline Risk Assessment in Superfund Remedy Selection Decisions.* OSWER Directive 9355.0-30. April 22, 1991.

————. 1993a. *Contract Laboratory Program National Functional Guidelines for Organic Data Review.* EPA-540/R-94/012.

————. 1993b. *Data Quality Objectives Process for Superfund.* EPA540-R-93-071. September 1993.

————. 1994a. *Test Methods for Evaluating Solid Waste, Physical/Chemical Methods.* SW-846.

————. 1994b. *Contract Laboratory Program National Functional Guidelines for Inorganic Data Review.* EPA-540/R-94/013.

URS Consultants, Inc. (URS). 1995a. *Quality Assurance Program Plan for the Comprehensive Long-Term Environmental Action Navy (CLEAN).*

————. 1995b. *Standard Operating Procedures for the Comprehensive Long-Term Environmental Action Navy (CLEAN).*

G. Fred Lee[1] and Anne Jones-Lee[1]

STORMWATER RUNOFF WATER QUALITY EVALUATION AND
MANAGEMENT PROGRAM FOR HAZARDOUS CHEMICAL SITES:
DEVELOPMENT ISSUES

REFERENCE: Lee, G. F. and Jones-Lee, A., **"Stormwater Runoff Water Quality Evaluation and Management Program for Hazardous Chemical Sites: Development Issues,"** *Superfund Risk Assessment in Soil Contamination Studies: Third Volume, ASTM STP 1338*, K. B. Hoddinott, Ed., American Society for Testing and Materials, 1998.

ABSTRACT: The deficiencies in the typical stormwater runoff water quality monitoring from hazardous chemical sites and an alternative approach (Evaluation Monitoring) for monitoring that shifts the monitoring program from periodic sampling and analysis of stormwater runoff for a suite of chemical parameters to examining the receiving waters to determine what, if any, water quality use impairments are occurring due to the runoff-associated constituents is presented in this paper. Rather than measuring potentially toxic constituents such as heavy metals in runoff, the monitoring program determines whether there is aquatic life toxicity in the receiving waters associated with the stormwater runoff. If toxicity is found, its cause is determined and the source of the constituents causing the toxicity is identified through forensic analysis. Based on this information, site-specific, technically valid stormwater runoff management programs can be developed that will control real water quality impacts caused by stormwater runoff-associated constituents.

KEYWORDS: hazardous chemical site, stormwater runoff monitoring, water quality impact evaluation

---

Introduction

Increasing attention is being given to managing the water quality impacts of stormwater runoff from "superfund" and other sites where hazardous chemicals exist in the near-surface or surface soils. Stormwater runoff from these areas could cause

---

[1]President and Vice-President, respectively, G. Fred Lee & Associates, 27298 E. El Macero, CA 95618-1005.

significant water quality impairment in off-site surface and groundwaters. This paper is a condensation of a more complete discussion of this topic. The more comprehensive review includes discussion of translocation as a source of stormwater runoff hazardous chemicals, cooperative watershed-based receiving water studies, aquatic sediment issues, atmospheric sources of toxic chemicals, importance of non-hazardous chemicals, wastewater discharges from site, groundwater recharge, data management and presentation, and duration of stormwater runoff monitoring. Information on these topics, as well as others, is available from Lee and Jones-Lee (1997a).

Deficiencies in Typical Stormwater Runoff Water Quality Monitoring Programs

In accord with US EPA stormwater runoff water quality regulatory requirements, the stormwater manager for the site is supposed to conduct analyses for any constituent that is likely to be present in stormwater runoff that could impair receiving water quality (US EPA 1990). For those sites complying with the minimum federal/state industrial site stormwater monitoring requirements, measurements are made of a stormwater sample's TDS, pH, TSS and TOC. For sites at which a wide variety of potentially hazardous chemicals have been manufactured, used, managed or disposed of, the stormwater runoff could justifiably be analyzed for the suite of Priority Pollutants.

Inappropriate Standards

At some hazardous chemical sites, drinking water quality is the focus of the site investigation where the stormwater runoff data are compared to drinking water MCLs in an approach similar to that followed in superfund site groundwater monitoring. For many of the heavy metals and some organics, the critical concentrations of constituents that are adverse to aquatic life are orders of magnitude lower than the concentrations that are acceptable in domestic water supplies. It is, therefore, important to consider the full range of potential impacts of stormwater runoff-associated constituents on the beneficial uses of the receiving waters for the stormwater runoff as part of developing a credible hazardous chemical site stormwater runoff water quality monitoring program.

Inappropriate Analytical Methods

One of the areas of particular concern in developing technically valid stormwater runoff water quality monitoring programs is the selection of analytical methods. At some sites, methods that are typically suitable for groundwater investigation are also used for surface water runoff. Such an approach can generate large amounts of "non-detect" data in which the detection limits are well above the potentially significant critical concentrations for adverse impacts of chemical constituents in the stormwater runoff to aquatic life in the receiving waters for the runoff.

An area that is often not adequately investigated in stormwater runoff from hazardous chemical sites is the potential for some of the runoff-associated constituents, such as chlorinated hydrocarbon pesticides, PCBs, dioxins and mercury, to bioaccumulate to excessive concentrations in the receiving water aquatic organisms to render these organisms unsuitable for use as human food because of an increased cancer risk or, in the case of mercury, neurological damage to those who consume the organisms. For many constituents, the concentrations in water that can lead to excessive bioaccumulation in fish are well below the analytical method detection limits typically used in "superfund" site investigations. This can lead to the superfund contractor incorrectly reporting that no water quality problems are associated with stormwater runoff from the site, since the concentrations of constituents found in the stormwater runoff are below the detection limits of the analytical methods used. Yet the fish in the waters receiving the stormwater runoff have bioaccumulated sufficient concentrations of hazardous chemicals derived from the site, as well as possibly elsewhere, to be hazardous for human consumption. These fish may also be hazardous for consumption by higher trophic level organisms such as fish-eating birds and mammals.

Even if appropriate analytical methods are used in measuring stormwater runoff-associated constituents from hazardous chemical sites and the data are properly compared to appropriate water quality criteria/standards designed to protect all beneficial uses such as fish and aquatic life, domestic water supply, agricultural water supply, etc., and an adequate sampling program has been conducted to measure the first-flush characteristics as well as the concentrations of constituents at other times during the runoff event for representative storms, it is still not possible to determine, from such monitoring programs what, if any, significant adverse impacts are occurring in the receiving waters for the stormwater runoff due to chemical constituents derived from the site in runoff waters.

Interpretation of Stormwater Runoff Data

The objective of a stormwater runoff water quality monitoring and evaluation program should be to determine whether the chemical constituents and/or pathogenic organisms in the runoff waters, when mixed into the receiving waters for the runoff, cause an impairment of the designated beneficial uses of the receiving waters, including downstream waters. The designated beneficial uses of concern are typically domestic water supply, fish and aquatic life, contact and other recreation, wildlife habitat, agricultural water supplies, etc. The typical approach used to determine whether stormwater runoff from a hazardous chemical or other site is adverse to the beneficial uses of the receiving waters involves comparing the concentrations of constituents found in the runoff waters to water quality criteria/standards. If an exceedance of the standard is found in the runoff waters, then it is often said that the stormwater runoff-associated constituent causing the exceedance is adverse to the beneficial uses of the receiving waters. However, in order to properly evaluate whether an exceedance of a water quality criterion/standard for a regulated chemical is adverse to fish and aquatic life in the receiving waters for the stormwater runoff, it is necessary to determine the concentrations of toxic/available forms of the constituent of concern in the receiving waters for the runoff at the point of mixing and downstream relative to the concentrations of this constituent that are known to be adverse to the forms of aquatic life present in, or that could be present in, receiving waters of concern for the site stormwater runoff waters. Also, the duration of exposure of the aquatic organism to toxic/available forms of the constituents in the runoff waters and in the receiving waters must be evaluated.

Use of US EPA Water Quality Criteria

As discussed by Lee and Jones (1991), Lee and Jones-Lee (1994a, 1995, 1996a,b, 1997b) and in references cited therein, the exceedance of a water quality criterion/ standard in stormwater runoff waters should not be interpreted to mean that a real water quality use impairment is occurring in the receiving waters for the runoff. Aquatic life-based water quality criteria/standards are typically developed based on worst-case, or near worst-case situations with respect to the constituent being adverse to the aquatic life. Normally, these criteria and standards assume that the constituents of concern are in 100% toxic/available forms and the potentially impacted organisms received extended-chronic exposures to these forms. The typical stormwater runoff event is normally of short duration relative to the critical duration-concentrations of toxic/available forms for aquatic life. Further, many of the chemical constituents in stormwater runoff are associated with particulates and are, therefore, in non-toxic, non-available forms. In some instances, the concentrations of constituents in stormwater runoff can be orders of magnitude above the water quality criterion/standard and not be adverse to the aquatic life-related beneficial uses of the receiving waters for the stormwater runoff. It cannot, however, be assumed that because this situation occurs at some locations that it will always occur at all locations and at all times. Site-specific investigations must be conducted to determine if the exceedance of a water quality standard represents a real water quality use impairment.

Under-Protective Nature of Some US EPA Water Quality Criteria

The authors have encountered a situation with chromium VI (Cr VI), where the US EPA water quality criterion of 10 µg/L would not necessarily be protective of aquatic life in receiving waters in which there is limited dilution of the stormwater runoff or wastewater discharges to a waterbody. Certain forms of aquatic life, such as zooplankton, which serve as important components of larval fish food, have been found to be adversely impacted by Cr VI at less than 0.5 µg/L. Lee and Jones-Lee (1997d) have reviewed the chromium chemistry/toxicity issues focusing on the deficiencies in the current regulatory approaches for control of chromium discharges to surface waters. Further, some regulatory agencies allow chromium III (Cr III) to be discharged in wastewaters and stormwaters at 50 µg/L, i.e. the chromium drinking water MCL. Such practice, however, can readily lead to chromium toxicity to aquatic life since Cr III can convert to Cr VI in some aquatic systems. It is important to understand that meeting the US EPA water quality criterion in ambient waters associated with stormwater runoff does not mean that the criterion constituent will not be adverse to some forms of aquatic life in the waterbody. In light of the information available today, it is appropriate to limit the total chromium concentration in a waterbody to 0.5 µg/L unless it can be shown that concentrations above this level are non-toxic to zooplankton, such as *Daphnia,* at the point of discharge for stormwater runoff and wastewater inputs, as well as downstream from this discharge/runoff.

There are a number of chemicals, such as arsenic, which are regulated as hazardous chemicals that will likely have their water quality standard significantly decreased within a few years. Arsenic has been used widely as a pesticide and herbicide. There are many former and current agricultural soils and some industrial areas that have sufficient arsenic in the surface soil to be of concern with respect to stormwater transport from the area. There is widespread recognition that arsenic at 50 µg/L (current drinking water MCL) represents a significant potential to cause cancer in people who consume domestic water supplies at or near these concentrations. The US EPA is reviewing the development of new, stricter drinking water standards for arsenic. The concentrations being considered are 0.2, 2.0, and 20 µg/L. It appears likely that a value of a few µg/L will be adopted, even though that value would still represent a significant cancer risk to those who consume waters with that concentration compared to the one in a million cancer risk that is typically accepted today as an appropriate risk for domestic water supplies.

The arsenic drinking water standard situation points to an important issue that needs to be considered in developing stormwater runoff water quality monitoring programs. Typically today, those establishing such programs only consider arsenic concentrations above 50 µg/L to be of potential concern with respect to water quality impacts. With new, stricter arsenic standards likely to be promulgated in the next few years, it is important to be certain that the analytical methods used are appropriate not only for today's regulatory situation but for those that can be reasonably expected to occur in the foreseeable future. Those conducting stormwater runoff monitoring programs from hazardous chemical sites should be cognizant of not only existing water quality criteria for water supplies, aquatic life, etc., but also of proposed changes such as will likely occur within a few years for arsenic or other constituents that have criteria currently under review. If, as in the case of arsenic, proposed lower concentration levels exist, then the stormwater managers should be using analytical methods that will measure the constituent at levels below the proposed criterion.

Unregulated Chemicals

One of the areas that should be of primary concern associated with stormwater runoff from a hazardous chemical site is the evaluation of the potential adverse impacts of the large number of potentially hazardous unregulated chemicals present in stormwater runoff. Typically, a comparison between the total organic carbon content of runoff waters and the total concentrations of specific organics measured in the runoff waters as determined by a Priority Pollutant scan shows that most of the organics in stormwater runoff are not identified/characterized. It is known that there are over 75,000 chemicals used in the US today. Only about 100 to 200 of these are regulated. Further, many chemicals can be transformed to other chemicals through chemical/biochemical processes that are also of concern with respect to impacting water quality as part of the unregulated chemicals in stormwater runoff. It is also known that each year newly developed and discovered hazardous chemicals are added to the list of hazardous chemicals that need to be regulated as a result of acquiring new information on their potential impacts. Therefore, it should never be assumed that because a stormwater runoff contains no constituents that cause an exceedance of a water quality criterion/standard in the runoff waters or receiving waters that the runoff-associated constituents from a hazardous chemical site or other area where complex mixtures of hazardous chemicals are present will not have an adverse impact on aquatic life and other beneficial uses of the receiving waters for the runoff.

A properly developed stormwater runoff impact evaluation and management program will include not only examination of the runoff waters and receiving waters for the regulated chemicals, but also include determining if the presence of unregulated, as well as regulated, chemicals in the runoff waters and the receiving waters are adversely impacting aquatic life and other beneficial uses of the receiving waters for the runoff. In evaluating the impact of stormwater runoff-associated constituents on receiving water water quality, it is important to examine the combined impacts of constituents in the stormwater runoff when mixed with constituents in the receiving waters. It is possible that adverse impacts will occur even though no impacts are potentially predicted based on examining the concentrations of regulated chemicals in the runoff waters. A combination of regulated and unregulated chemicals in the runoff waters and receiving waters could have an adverse impact that would not occur in either water alone. As discussed in a subsequent section of this paper, toxicity testing of discharge in ambient waters can be used to screen for combined effects of potentially toxic regulated and unregulated chemicals on receiving water aquatic life.

In addition, it is necessary to consider not only potential impacts at the point of mixing of the runoff waters with the receiving waters, but also downstream of this point where, associated with chemical/biochemical transformations, regulated and unregulated hazardous chemicals and non-hazardous chemicals are converted to hazardous forms. An example of this type of situation occurs with Cr III which some regulatory agencies allow to be discharged at 50 μg/L based on toxicity to humans. Generally, because of its low level of toxicity to aquatic life, Cr III is regulated based on drinking water standards of 50 μg/L. However, since Cr III can convert to Cr VI in aerobic surface waters, which can be toxic at about 0.5 μg/L, the discharge of Cr III at 50 μg/L, which is non-toxic at the point of discharge, can lead to aquatic life toxicity in the receiving waters downstream of the discharge due to the conversion of Cr III to Cr VI. As discussed by Lee and Jones-Lee (1997d). This conversion can take place in a few hours or over several days, depending primarily on receiving water conditions.

Another issue of concern with respect to the discharge of Cr III to a waterbody is the accumulation of the Cr III in the waterbody sediments. Cr III has a strong tendency to sorb to particulates and precipitate. This tends to cause Cr III particulate forms to accumulate in receiving water sediments during low-flow conditions. These areas of accumulation can, however, be scoured into the water column during high-flow conditions, suspending the particulate Cr III into the water column. This could cause significantly elevated levels of Cr III to pass downstream during elevated flow conditions. It was recently reported by Gunther (1997) that elevated flow conditions have apparently scoured Cr III from sediments in the Sacramento River system and have bioaccumulated in mussels in San Francisco Bay. This situation raises the question of whether the scoured Cr III, either through bioaccumulation or through conversion to Cr VI, is adverse to the beneficial uses of downstream waters where Cr III has accumulated.

The Cr III discharge situation provides another example of the inappropriateness of assuming that just because stormwater runoff contains constituents that, in the runoff or at the point of mixing with the receiving waters, are not adverse to aquatic life, that these constituents will not be adverse to aquatic life at some time in the future under a different flow regime. It is essential that receiving water studies be conducted to determine whether adverse conditions are found in the receiving waters due to stormwater runoff-derived constituents at the time of discharge as well as in the future under different flow or other conditions that can exist in the receiving waters.

Regulatory agencies and dischargers often use an arbitrary fixed distance (such as 50 or 100 meters) for sampling the receiving waters for discharges. The downstream sampling station should be selected based on a site-specific evaluation of mixing distances at various receiving water flows. Since the distance for mixing is dependent upon the receiving water velocity, consideration must be given to how the velocity of the receiving waters changes as a function of flow in selecting downstream sampling stations. Further, the rates of reactions of potential concern, such as the conversion of Cr III to Cr VI, must be considered.

An Alternative Monitoring/Evaluation Approach

In order to be protective of aquatic life and other beneficial uses of receiving waters for stormwater runoff, it should be assumed that the exceedance of a water quality criterion/standard due to runoff-associated constituents from a hazardous chemical site represents an adverse impact that should be evaluated by the stormwater manager to determine if the potential impact is, in fact, manifested in the receiving waters for the runoff. Lee and Jones-Lee (1996a,b, 1997b,c) have developed an Evaluation Monitoring (EM) approach that can be used to determine whether exceedances of water quality criteria/standards in stormwater runoff are causing real water quality use impairments in the receiving waters for the runoff. This approach can also detect some water quality problems due to unregulated chemicals. EM focuses on finding a real water quality problem-use impairment in the receiving waters for the stormwater runoff, identifying its cause and determining whether stormwater runoff-associated constituents derived from a particular site cause a water quality use impairment(s) in the receiving waters for the runoff. It should not be assumed, as it is often done, that an exceedance of a water quality standard represents a real water quality use impairment. Exceedances can readily reflect the overly-protective nature of many water quality criteria/standards which fail to consider the toxic, available forms of the constituents in runoff waters as well as the duration of exposure to excessive concentrations of available forms of constituents in the receiving waters.

The EM approach should be implemented as a watershed-based water quality management program in which all stakeholders (dischargers, regulatory agencies and potentially impacted parties) work together to define whether stormwater runoff from a particular location causes real, significant water quality problems in the receiving waters for the runoff. Where such problems are found, the stakeholders work together to control them. All of the designated beneficial uses for the receiving waters are considered in implementing the EM approach. These may include domestic water supply, fish and aquatic life, public health, contact and other recreation, wildlife habitat, agricultural water supply, groundwater recharge, excessive fertilization, etc. Presented below is a summary of some of the issues that need to be considered in implementing the EM approach to determine whether stormwater runoff from a hazardous chemical site is adversely impacting some of the beneficial uses of receiving waters for the runoff. Additional information on this topic is provided in the references listed for this paper.

Aquatic Life Toxicity

There are basically two concerns for the protection of aquatic life and their use as food associated with the discharge of hazardous/deleterious chemicals from hazardous chemical sites and other areas where there are complex mixtures of regulated and unregulated chemicals. One of these is aquatic life toxicity; the other is excessive bioaccumulation. Lee and Jones (1991) and Lee and Jones-Lee (1994a,1996a,b,c) recommend that all stormwater runoff water quality evaluation programs include measurement of the aquatic life toxicity to sensitive forms of aquatic life in the stormwater runoff waters as well as the receiving waters mixed with the stormwater runoff near the point of mixing and downstream thereof. The chronic toxicity endpoint should be used.

Appropriately conducted aquatic life toxicity tests can screen stormwater runoff waters for potential adverse impacts of all regulated and unregulated chemical constituents that are of concern with respect to being potentially toxic to aquatic life in the receiving waters. Such tests can and usually show that exceedances of water quality criteria for potentially toxic regulated chemicals in stormwater runoff do not necessarily translate to toxicity in the receiving waters for the runoff that is adverse to aquatic life in these waters. Further, such testing, if appropriately conducted, can detect toxic components in the unregulated chemicals in the stormwater runoff. Aquatic life toxicity testing using standardized US EPA toxicity test methods for freshwater systems using fathead minnow larvae and *Ceriodaphnia* (Lewis et al.1994) is a powerful tool that should be used on a routine basis to determine whether potentially toxic regulated and unregulated chemicals in stormwater runoff cause significant toxicity in receiving waters for the runoff.

Toxicity testing of the stormwater runoff waters should be done for at least two storms each season. In addition to testing the runoff waters for aquatic life toxicity, the receiving waters should be sampled and tested for aquatic life toxicity just downstream of where the runoff waters enter and are mixed with the receiving waters. The degree of mixing should be established at that point by measurements of specific conductance and/ or temperature in the receiving waters. Further, toxicity testing should be done further downstream to detect whether non-toxic constituents in the runoff waters convert to toxic constituents in the receiving waters in sufficient amounts to be toxic to aquatic life, i.e. Cr III to Cr VI conversion.

It was through the use of aquatic life toxicity testing of stormwater runoff that diazinon and chlorpyrifos, organophosphate pesticides that are widely used in urban areas and in agriculture, were found to cause potentially significant aquatic life toxicity due to their presence in stormwater runoff from urban areas, highways and some rural areas.

Lee and Taylor (1997) have recently reviewed the stormwater runoff organophosphate toxicity issue. Diazinon is one of the unregulated chemicals that is causing widespread aquatic life toxicity in receiving waters for stormwater runoff. There are periods of time for several weeks each year in north-central California where diazinon applied to orchards as a dormant spray causes significant aquatic life toxicity in all runoff waters from urban and rural areas. A significant portion of this diazinon is volatilized at the time of application and transported through the atmosphere. It becomes surface water runoff through atmospheric scouring associated with rainfall and fogfall. Similarly, urban area stormwater runoff has been found to contain diazinon and chlorpyrifos toxicity due to homeowner structural and landscape use of these chemicals. There are other organophosphate and carbamate pesticides that are not now being adequately regulated which are likely causing similar problems in stormwater runoff from urban and rural areas.

It is important, when using aquatic life toxicity testing, to properly interpret the results of such tests. It should not be assumed that because aquatic life toxicity is found in the stormwater runoff waters that this will lead to significant aquatic life toxicity in the receiving waters for the runoff. As discussed by Lee and Jones (1991) and Lee and Jones-Lee (1994b, 1996c), the short-term episodic nature of stormwater runoff events, coupled with the approach used in aquatic life toxicity testing where the tests are conducted for approximately one week, can over-estimate toxicity that occurs in the receiving waters due to the discharge of toxic stormwater runoff. Typical aquatic life toxicity testing requires several days to about a week duration of test organism exposure. It is unusual in a stormwater runoff event for aquatic organisms to receive a week-long duration of exposure to toxic conditions. There are, however, significant differences in the rate at which various types of chemicals exert a toxic effect. There are fast-acting chemicals where, at elevated concentrations, the toxicity can be manifested within a few hours. There are other chemicals where, either due to low concentrations or the typical rate at which toxicity is manifested, several days of exposure must occur before the organisms are adversely impacted.

The recommended approach is to use the standard toxicity test to screen stormwater runoff for potential adverse impacts associated with the discharge of potentially toxic regulated and unregulated chemicals in the stormwater runoff. If toxicity is found in the runoff waters that persists for a sufficient period of time in the receiving waters to be potentially adverse to aquatic life, then additional toxicity testing should be conducted in which the toxicity-duration of exposure relationship that occurs in the receiving waters for the runoff is mimicked in the toxicity test. Lee and Jones (1991) and Lee and Taylor (1997) provide additional information on this topic.

Bioaccumulation

The bioaccumulation of hazardous chemicals in aquatic life tissue is one of the major adverse impacts that can occur due to stormwater runoff constituents. The chemicals of greatest concern for excessive bioaccumulation are the chlorinated hydrocarbon pesticides such as DDT and chlordane, PCB's, dioxins, and mercury. These chemicals have been found to bioaccumulate to a sufficient extent in aquatic life tissue to cause the use of aquatic organisms as food to be hazardous due to increased cancer risk or other adverse impacts on human health. For many years, the Food and Drug Administration (FDA) Action Levels were used to determine excessive concentrations of hazardous chemicals in aquatic life tissue. The US EPA (1994) adopted risk-based tissue concentrations which consider the potential hazard that a particular concentration of a hazardous chemical in an organism tissue represents through its cancer potency as well as the amount of tissue consumed. Generally, the risk-based excessive concentrations of hazardous chemicals are one or more orders of magnitude lower than the FDA Action Levels. This has created a situation where concentrations of hazardous chemicals in stormwater runoff that were not considered to have adverse impacts in the past are now being recognized as a potential significant source of excessive concentrations of the chemicals in fish tissue. Lee and Jones-Lee (1996b,d) have provided guidance on assessing excessive bioaccumulation of chemicals in aquatic organisms associated with stormwater runoff situations.

Similar problems occur with mercury where mercury is determined with analytical methods that do not have the necessary sensitivity to determine whether, under worst-case conditions such as those used by the US EPA in developing bioaccumulation-based water quality criteria, measured concentrations could result in excessive bioaccumulation in the receiving water organisms. The technically valid, cost-effective approach for assessing excessive bioaccumulation is a direct measurement of edible aquatic organism tissue residues. This is the approach that Lee and Jones-Lee (1996a,b,d) recommend as part of implementing the EM approach for stormwater runoff. Direct measurements of excessive bioaccumulation can readily be accomplished with the analytical methods available today. Such measurements provide information that can be used to determine whether all sources of a bioaccumulatable chemical in fish tissue contributes sufficient amounts of the chemical to the waterbody in available forms to lead to a bioaccumulation-caused use impairment of the waterbody. If the concentrations of chlordane, mercury or some other constituent are below excessive levels within aquatic organisms taken within the waterbody and downstream of the point at which the stormwater runoff enters the waterbody, then it can be concluded that excessive bioaccumulation is not occurring and, most importantly, that the stormwater runoff does not contain sufficient concentrations of a potential bioaccumulatable chemical to cause a water quality use impairment in the receiving waters due to that chemical.

If, however, the fish and/or other aquatic organisms used as human food in the receiving waters potentially impacted by stormwater runoff from a hazardous chemical site contain excessive concentrations of a chemical such as chlordane or mercury, then it is necessary to conduct additional studies to determine whether the stormwater runoff is a significant contributor of the constituent of concern to cause or contribute to the excessive bioaccumulation problem. Lee and Jones-Lee (1996a,b 1997b) have discussed the use of forensic procedures with caged organisms and/or laboratory studies that can determine whether stormwater runoff-associated potentially bioaccumulatable constituents could be significant contributors to excessive bioaccumulation in receiving water aquatic organisms. The bioaccumulation studies of stormwater derived from hazardous chemical sites should involve measurement of receiving water aquatic organism tissue levels for the conventional suite of potentially significant bioaccumulatable chemicals, such as the chlorinated hydrocarbon pesticides, PCBs, dioxins and mercury, for a several-year period. The bioaccumulation studies should be conducted each spring and fall to examine seasonal differences. Higher trophic level predator organisms should be sampled as well as the organisms that tend to have higher concentrations of fat in their tissue. Generally, the chlorinated hydrocarbon pesticides and PCBs accumulate in high lipid content tissue to a greater degree. The lipid content of the tissues collected and analyzed should also be determined.

Caution must be exercised in using US EPA water quality criteria, such as the US EPA (1987) "Gold Book" criteria, in predicting bioaccumulation that will occur in receiving waters for stormwater runoff. The US EPA water quality criteria for bioaccumulation are based on worst-case conditions which tend to over-estimate the actual bioaccumulation that will occur in many waterbodies. Lee and Jones-Lee (1996d) have discussed the current information in using chemical concentration data in water and/or sediments to estimate excessive bioaccumulation that occurs in receiving water aquatic organisms. There are a wide variety of factors which influence whether a particular chemical constituent, as typically measured by standard water quality monitoring analytical procedures, will bioaccumulate to excessive levels. The most important factor is the aqueous environmental chemistry of the constituent. Many of the potentially bioaccumulatable chemicals exist in a variety of chemical forms which are non-available to bioaccumulate within aquatic organisms. Since the analytical methods typically used in water quality investigations rarely only measure available forms of constituents, most measurements of bioaccumulatable chemicals tend to over-estimate the actual bioaccumulation that will occur when the concentrations are used with US EPA bioaccumulation factors. It is for this reason that the primary tool for determining whether excessive bioaccumulation occurs in a waterbody due to stormwater runoff-associated constituents is the actual bioaccumulation in aquatic organisms in the receiving waters. This is a more reliable approach than the approach that is typically used today of trying to measure concentrations of bioaccumulatable chemicals in runoff waters and then extrapolating these concentrations to excessive concentrations in receiving water aquatic organisms.

Aquatic Sediment Issues

The transport of hazardous chemicals from superfund and other hazardous chemical sites occurs with dissolved chemicals and chemicals attached to particulate matter. The dissolved chemicals can interact with the receiving water particulates to become part of the particulate-associated hazardous chemicals derived from the site. Since the particulate-associated chemicals are transported differently and represent significantly different hazards in the environment, it is important to determine whether hazardous conditions exist in the receiving waters due to the release of hazardous chemicals from the site that are in particulate forms or become particulate in the receiving waters.

Generally, particulate forms of hazardous chemicals are non-toxic and non-available and therefore represent minimal hazards in the environment. There are situations, however, where the accumulation of particulate forms in bedded sediments represents a potential cause of water quality deterioration due to either aquatic life toxicity to benthic or epibenthic organisms, or serve as a source of bioaccumulatable chemicals that can be adverse to the beneficial uses of a waterbody through causing a health hazard to humans who use the aquatic life as food. There is also the potential for the bioaccumulation of hazardous chemicals to be adverse to higher trophic level organisms, such as fish-eating birds and terrestrial mammals.

The US EPA (1995) has officially recognized that particulate forms of many heavy metals are non-toxic and non-available and now recommends regulating these heavy metals based on ambient water dissolved forms. The Agency should adopt the same approach for many of the potentially toxic organics and other constituents that tend to become associated with particulates. The Agency still recommends measuring total recoverable metals in discharge waters and the use of a generic or site-specific translator to translate dissolved forms of metals to particulate forms and vice versa in the ambient waters receiving the heavy metal input. This is based on the concern that particulate forms present in the discharge would convert to dissolved forms in the receiving waters. It is the authors' experience that such a conversion would be extremely rare.

Manuscript page limitations prevent the inclusion of a discussion of receiving water bedded sediment water quality issues. Such issues are important at many hazardous chemical sites. For a discussion of these issues, consult Lee and Jones-Lee (1997a).

Overview of Recommended Stormwater Monitoring for Hazardous Chemical Sites

Typically, stormwater runoff from urban areas and highways has been found to have limited adverse impact on the beneficial uses of the receiving waters for stormwater runoff (Lee and Jones-Lee 1996e). However, this is not necessarily the situation for stormwater runoff from hazardous chemical sites or other areas where large amounts of potentially hazardous chemicals are used in such a way as to possibly be present in elevated concentrations in stormwater runoff from the area. Sites of this type deserve special monitoring and proper interpretation of the US EPA's General Industrial Permit requirements for monitoring for "toxic" chemicals that could be present in the stormwater runoff. There is growing recognition that conventional end-of-the-pipe/edge-of-the-pavement/property monitoring of stormwater runoff for the conventional as well as the Priority Pollutants provides limited information on the impact of the stormwater runoff on receiving water quality-beneficial uses (Lee and Jones-Lee 1994a,b, 1996a,b,c, 1997b,c). There is widespread recognition that the conventional monitoring approach for stormwater runoff needs to be shifted from runoff water monitoring to receiving water monitoring and evaluation. The EM approach, in which the regulated entity (the PRP for a hazardous chemical site under remediation), regulatory agencies and the impacted community work together in a watershed-based water quality evaluation and management program to define what, if any, real water quality use impairments are occurring in the receiving waters for the stormwater runoff, if implemented properly, can be a technically valid, cost-effective approach. A key component of an appropriate stormwater runoff monitoring program is the examination of the runoff waters and receiving waters for aquatic life toxicity and excessive bioaccumulation of hazardous chemicals that cause receiving water organisms to be considered hazardous for use as human food.

It is important in developing a stormwater runoff water quality evaluation and management program for a hazardous chemical site to not over-regulate stormwater discharges and thereby waste public and private funds in unnecessary monitoring. In situations where adequate monitoring and evaluation have been conducted which show with a high degree of certainty that there is limited likelihood of significant adverse impacts on the beneficial uses of the receiving waters for the runoff, there is no point in continuing intensive monitoring and management programs. A low-level, on-going monitoring program should be continued in order to be certain that new problems do not occur in the future that were not detected previously or that the site characteristics changed sufficiently to significantly change the concentrations of constituents in the stormwater runoff.

The overall approach that should be used in a monitoring and management program of hazardous chemical sites is to err on the side of public health and environmental protection in those situations where definitive information on the impact of runoff from a site is lacking. Regulatory agencies should require that the burden of proof should be on the PRP - stormwater discharger to reliably demonstrate that stormwater runoff from the site is not adverse to the beneficial uses of the receiving waters.

Conclusions

The development of a technically valid, cost-effective stormwater runoff monitoring and evaluation program for a hazardous chemical site requires a high degree of understanding of aquatic chemistry, aquatic toxicology, hydrodynamics and water quality. The US EPA's General Industrial Permit stormwater runoff monitoring program will not ordinarily provide adequate monitoring to ensure that hazardous chemical site stormwater runoff-associated constituents do not have an adverse impact on the beneficial uses of the receiving waters for the runoff. Credible stormwater runoff monitoring programs must involve in-depth, reliable examination of the water quality characteristics of the receiving waters for the runoff. The EM approach provides a focused examination of receiving waters in which these waters are examined for water quality use impairments of potential concern to the public and others who utilize these waters. If properly implemented, the EM approach can significantly reduce the cost of monitoring of hazardous chemical sites' stormwater runoff and focus the funds spent on monitoring on detecting real water quality problems that need to be addressed in order to protect the designated beneficial uses of the receiving waters for the stormwater runoff.

References

Gunther, A., February 1997, "Bivalve Bioaccumulation," Presentation at San Francisco Bay Regional Monitoring Program Annual Meeting, Oakland, CA.

Lee, G. F. and Jones, R.A., 1991, "Suggested Approach for Assessing Water Quality Impacts of Urban Stormwater Drainage," *Symposium Proceedings on Urban Hydrology*, American Water Resources Association, AWRA Technical Publication Series TPS-91-4, AWRA, Bethesda, MD, pp. 139-151.

Lee, G. F. and Jones-Lee, A., 1994a, "Deficiencies in Stormwater Quality Monitoring," *Proc. Engineering Foundation Conference*, American Society of Civil Engineers, New York, NY, pp. 651-662.

Lee, G. F. and Jones-Lee, A., 1994b, "Stormwater Runoff Management: Are Real Water Quality Problems Being Addressed by Current Structural Best Management Practices? Part 1," *Public Works*, Vol.125, pp. 53-57, 70-72; Part Two, 1995, Vol 126, pp.54-56.

Lee, G. F. and Jones-Lee, A., 1995,"Appropriate Use of Numeric Chemical Water Quality Criteria," *Health and Ecological Risk Assessment*, Vol 1, pp.5-11. Letter to the Editor, Supplemental Discussion, 1996, Vol 2, pp.233-234.

Lee, G. F. and Jones-Lee, A., 1996a, "Assessing Water Quality Impacts of Stormwater Runoff," North American Water & Environment Congress, Published on CD-ROM, Amer. Soc. Civil Engr., New York, NY. Available at: http://members.aol.com/gfredlee/gfl.htm.

Lee, G. F. and Jones-Lee, A., 1996b, "Evaluation Monitoring as an Alternative for Conventional Stormwater Runoff Water Quality Monitoring and BMP Development," Report of G. Fred Lee & Associates, El Macero, CA.

Lee, G. F. and Jones-Lee, A., 1996c, "Stormwater Runoff Quality Monitoring: Chemical Constituent vs. Water Quality, Part I, II," *Public Works*, Part I: Vol. 147, pp. 50-53; Part II: December.

Lee, G. F. and Jones-Lee, A., 1996d, "Summary of Issues Pertinent to Regulating Bioaccumulatable Chemicals," Report G. Fred Lee & Associates, El Macero, CA.

Lee, G. F. and Jones-Lee, A., 1996e, "Results of Survey on Water Quality Problems Caused by Urban and Highway Stormwater Runoff," *Runoff Reports*, Vol. 4, No. 5, pp. 3.

Lee, G. F. and Jones-Lee, A., April 1997a, "Development of a Stormwater Runoff Water Quality Evaluation and management Program for Hazardous Chemical Sites: UCD-DOE LEHR Superfund Site Experience," Report, G. Fred Lee & Associates, El Macero, CA.

Lee, G. F. and Jones-Lee, A., 1997b "Development and Implementation of Evaluation Monitoring for Stormwater Runoff Water Quality Impact Assessment and Management," Report of G. Fred Lee & Associates, El Macero, CA.

Lee, G. F. and Jones-Lee, A., 1997c, "Evaluation Monitoring as an Alternative to Conventional Stormwater Runoff Monitoring and BMP Development," *SETAC News*, Vol. 17, No. 2, pp. 20-21.

Lee, G. F. and Jones-Lee, A., April 1997d, "Chromium Speciation: Key to Reliable Control of Chromium Toxicity to Aquatic Life," Presented at the American Chemical Society National Meeting poster session, San Francisco, CA. Poster items available from: http://members.aol.com/gfredlee/gfl.htm.

Lee, G. F. and Taylor, S., 1997, "Aquatic Life Toxicity in Stormwater Runoff to Upper Newport Bay, Orange County, California: Initial Results," Report of G. Fred Lee & Associates, El Macero, CA.

Lewis, P. A., Klemm, D. J., Lazorchak, J. M., Norberg-King, T. J., Peltier, W. H., Heber, M. A., 1994, *Short-Term Methods for Estimating the Chronic Toxicity of Effluents and Receiving Waters to Freshwater Organisms*, Third Edition, Environmental Monitoring Systems Laboratory, Cincinnati, OH, Environmental Research Laboratory, Duluth, MN, Region 4, Environmental Services Division, Athens, GA, Office of Water, Washington, D.C., Environmental Monitoring Systems Laboratory-Cincinnati, Office of Research and Development, US Environmental Protection Agency, Cincinnati, OH.

US EPA, 1987, *Quality Criteria for Water 1986*," US Environmental Protection Agency, Office of Water Regulations and Standards, EPA 440/5-86-001, Washington, D.C.

US EPA, 1990, "National Pollutant Discharge Elimination System Permit Application Regulations for Stormwater Discharges; Final Rule," 40 CFR Parts 122, 123 and 124, *Federal Register*, Vol. 55, No. 222, pp. 47990-48901, November 16.

US EPA, 1994, *Guidance for Assessing Chemical Contaminant Data for Use in Fish Advisories, Vol. II, Risk Assessment and Fish Consumption Limits*, US EPA Office of Water, EPA 823-B-94-004, Washington, D.C.

US EPA, 1995, "Stay of Federal Water Quality Criteria for Metals; Water Quality Standards; Establishment of Numeric Criteria for Priority Toxic Pollutants; States' Compliance--Revision of Metals Criteria; Final Rules," *Federal Register*, Vol. 60, No. 86, pp. 22228-22237.

Copies of the authors' reports which serve as background to this paper are available from their web site (http://members.aol.com/gfredlee/gfl.htm).

D. Wayne Berman[1], Bruce C. Allen[2], and Cynthia B. Van Landingham[3]

EVALUATION OF THE PERFORMANCE OF STATISTICAL TESTS USED IN
MAKING CLEANUP DECISIONS AT SUPERFUND SITES. PART 1:
CHOOSING AN APPROPRIATE STATISTICAL TEST

REFERENCE: Berman, D.W., Allen, B.C., and Van Landingham, C.B., **"Evaluation of
the Performance of Statistical Tests Used in Making Cleanup Decisions At
Superfund Sites. Part 1: Choosing an Appropriate Statistical Test,"** *Superfund Risk
Assessment in Soil Contamination Studies: Third Volume, ASTM STP 1338*, K. B.
Hoddinott, Ed., American Society for Testing and Materials, 1998.

ABSTRACT: The decision rules commonly employed to determine the need for cleanup
(including those cited in recent guidance) are evaluated below both to identify conditions
under which they lead to erroneous conclusions and to quantify the rate that such errors
occur. Their performance is also compared with that of other applicable decision rules.

We based the evaluation of decision rules on simulations. Results are presented as
power curves. These curves demonstrate that the degree of statistical control achieved is
independent of the form of the null hypothesis. The loss of statistical control that occurs
when a decision rule is applied to a data set that does not satisfy the rule's validity criteria
is also clearly demonstrated. Some of the rules evaluated do not offer the formal
statistical control that is an inherent design feature of other rules. Nevertheless, results
indicate that such "informal" decision rules may provide superior overall control of error
rates, when their application is restricted to data exhibiting particular characteristics.

The results reported here are limited to decision rules applied to uncensored and
lognormally distributed data. To optimize decision rules, it is necessary to evaluate their
behavior when applied to data exhibiting a range of characteristics that bracket those
common to field data. The performance of decision rules applied to data sets exhibiting a
broader range of characteristics is reported in the second paper of this study.

KEYWORDS: risk assessment, site assessment, statistics, decision error, error rates,
hypothesis tests, confidence limits

---

[1]President, Aeolus Environmental Services, 751 Taft St., Albany, CA 94706.

[2]Project Manager, ICF Kaiser Engineers, P.O. Box 14348, Research Triangle Park,
NC, 27709.

[3]Project Manager, ICF Kaiser Engineers, 602 East Georgia, Ruston, LA 71270.

This is the first of two papers in which decision rules commonly employed to determine the need for cleanup at Superfund sites are evaluated both to identify the conditions under which they may lead to erroneous conclusions and to quantify the rate at which such errors may occur. The performance of these procedures (in terms of their error rates) is also compared with the performance of other decision rules that might alternately be employed to determine whether cleanup is warranted.

In this first paper, considerations that need to be addressed to evaluate performance are introduced and performance is judged for cases in which the procedures are applied to data exhibiting reasonably ideal behavior. In the second paper (Berman et al. 1998), performance is evaluated for cases in which the various decision rules are applied to data exhibiting a range of characteristics that more closely mimic what is observed among environmental data.

Background

In most site risk assessments, the decision whether to remediate ultimately reduces to determining whether the arithmetic mean, "m," of some contaminant concentration within a defined area or volume element of the environment exceeds a critical target value, "Q" (USEPA 1992). In this context, m is the *true* arithmetic mean of the contaminant concentration and Q typically represents the boundary between acceptable and unacceptable conditions (i.e., the boundary between acceptable concentrations and those that are large enough to present unacceptable risks). The manner in which such comparisons fit into the larger scheme of a risk assessment has been described previously (Berman 1995).

The comparison between Q and estimates of m represents a test in which a *null* hypothesis (e.g. that m < Q, which might indicate that a site is "clean") is assumed to be true unless conditions at the site warrant rejecting the null in favor of an alternate hypothesis (e.g. that m ≥ Q, which might indicate that cleanup is required). It is equally valid to begin with the assumption that the site is "contaminated" (i.e. that m > Q) and to test whether this null can be rejected in favor of an alternate (i.e. that m ≤ Q) to "prove" that a site is clean. In fact, current guidance favors this latter formulation when making decisions at Superfund sites (USEPA 1992). As shown below, however, these two hypothesis formulations are equivalent so that, when incorporated into any particular decision rule, the degree of statistical control that is achieved is the same for both formulations.

In the absence of error (uncertainty and variability) in an estimate of m, a direct comparison between m and Q provides a perfect test of the null hypothesis. However, the true value of m can never be known exactly. This is due primarily to the limited ability of a finite number of samples that are collected from a large and variable environmental medium to completely represent the characteristics of that medium. Also, methods employed for sampling and analysis are subject to error and this introduces additional uncertainty in the estimates of m. Thus, summary statistics that are biased estimates of m (such as upper or lower confidence limits) are typically derived from field measurements and compared to Q to determine *with some predefined level of confidence* whether m is likely to exceed Q. In fact, a variety of summary statistics can be (and have been) used to test hypotheses concerning the true mean of a field distribution of concentrations and each offers various advantages and limitations.

Note that the degree to which a set of samples can be assumed to represent the sampled medium is a strong function of the process by which locations were selected for the samples collected (which affects the representativeness of the data), the number of samples collected, the characteristics of the methods employed for sampling and analysis, and the homogeneity of the data. The effects on decision errors associated with the data quality limitations imposed by each of these factors were explored in a previous study (Berman 1995). In the current study, we assumed that optimal data are available for evaluating the site so that the performance of the various decision rules themselves can be investigated.

Decision Rules

The nature of the summary statistic selected to represent the mean, m, the manner in which it is calculated from field measurements, and the formulation of the hypotheses employed to compare that statistic to the target, Q, constitute a "decision rule." Decision rules evaluated in this study include those using as summary statistics: upper or lower confidence limits to the mean, best estimates of the mean, the maximum measured concentration, or the 90th and 95th percentiles of the sampled distribution.

Lower and Upper confidence limits to the mean based both on the t-statistic, which is designed for normally distributed data (Bickel and Doksum 1977), and on the Land Equation, which is designed for lognormally distributed data (Land 1971), are each evaluated in this study. The former were used in the early days of the Superfund program and are still used under certain circumstances. The latter are recommended in more recent guidance (e.g. USEPA 1994).

Best estimates of the mean evaluated in this study are based both on the common estimator, the arithmetic average, which is an unbiased estimator of the mean of *any* distribution (e.g. Box et al. 1978) and on the simplified maximum likelihood estimator for the mean of a lognormal distribution (Gilbert 1987). While not designed to provide formal statistical control of error (as described in the next section), these have been used to evaluate data in screening procedures.

Performance and Error Rates

The probability that a decision rule leads to an incorrect decision (i.e. the error rate) is a function both of the characteristics of the decision rule and of the underlying true condition at the site. Given a particular decision rule in which, for example, it is assumed that m < Q as a null hypothesis and a comparison between Q and some summary statistic employed to represent m is used to decide whether to accept or reject the null hypothesis, there are four possible outcomes:

- m is truly less than Q and the comparison leads properly to a conclusion that m is less than Q;

- m is truly less than Q but the comparison leads falsely to a conclusion that m exceeds Q;
- m truly exceeds Q and the comparison leads properly to a conclusion that m exceeds Q; or
- m truly exceeds Q but the comparison leads falsely to a conclusion that m does not exceed Q.

The first and third of the above listed outcomes are correct results. The second and fourth are erroneous. Thus, there are two kinds of errors that can potentially occur during the application of a decision rule. A "Type I" error occurs when the null hypothesis is falsely rejected. In the illustration presented above, this corresponds to the second of the listed outcomes. A "Type II" error occurs when the null hypothesis is falsely accepted and corresponds to the last of the possible outcomes listed in the above illustration.

The relationship between Type I and Type II errors and the specific outcomes of a comparison depends on the way that the null hypothesis is defined. Thus, for example, if a new decision rule is examined in which the null hypothesis is assumed to be $m > Q$ (the reverse of the illustration above), then the second of the four outcomes listed above becomes a Type II error and the fourth becomes a Type I error.

In the face of uncertainty, as indicated previously, formal decision rules are applied to control decision error. Statistical control is achieved when, due to the inherent design of a particular statistical test, the null hypothesis is rejected at a *fixed and defined* rate at the decision point (i.e. when $m = Q$). The rate that the null is rejected at the decision point is termed the significance or α (alpha) level for the test.

Importantly, statistical control implies that the significance level of the test is independent of sample size and other characteristics of the data to which the decision rule is applied. However, the rate of rejection of the null hypothesis for true conditions other than the decision point (i.e. when $m \neq Q$), is generally a function of both the true condition (i.e. the value of m) and the sample size. The rate at which the null hypothesis is rejected for conditions under which rejection is correct (i.e. when $m \geq Q$) is termed the power of the test. If a statistical test is well behaved, the power of the test increases as sample size increases.

In practice, the intended statistical control of a particular test is achieved when the characteristics of the data to which the test is applied completely satisfy the validity criteria for the test. The validity criteria for a particular statistical test are the mathematical axioms that were employed to construct the test (e.g. some tests require that data distributions be symmetric). However, tests are frequently applied to data that exhibit characteristics that do not strictly satisfy such criteria. If deviations are small, the test may still behave reasonably and statistical control will be maintained. However, the characteristics of environmental data frequently fall at the fringe of acceptability for the kinds of spatially-independent statistical tests that are commonly applied.

Environmental Data

Depending on how environmental samples are collected and depending on what is being sampled, the distribution of concentrations being evaluated may be quite irregular (e.g. Berman 1995). It is even possible, for example, that such distributions are multi-modal. With proper planning (provided that certain information is initially available), it may be possible to avoid many of the more extreme departures from well-behaved data sets. However, the required information may not always be available.

As an illustration, consider the common case of sampling an area where the precise boundaries of contamination are unknown. In this case, one may be sampling a mix of contaminated and uncontaminated areas so that a bi-modal distribution results, which is composed of some distribution of contaminant concentrations combined with a finite density of concentrations precisely at zero (or some level that is characteristic of natural background). Of course the precise nature of this distribution may also be masked by censoring, which occurs because it is impossible to obtain concentration information below the detection limit of the analytical method.

Even under the best of circumstances, environmental data are constrained to be non-negative, their distributions tend to be asymmetric and positively skewed (i.e. means tend to be greater than medians), and they tend to exhibit substantial spread (i.e. variability). One measure of spread or variability, the coefficient of variation (CV), is defined as the standard deviation divided by the mean. CV's that are commonly observed among environmental data tend to exceed one and are frequently even larger. In fact, CV's as large as 5 or 10 are not uncommon, although most data sets exhibit CV's ranging up to about 5 (Berman 1995).

The Problem

When statistical tests are applied to environmental data exhibiting characteristics that do not satisfy the validity criteria for these tests, the intended statistical control may be compromised. Consequently, the error rates for decisions based on such applications may be larger than that expected from the "advertised" performance of the test. In this paper, performance is measured and reported for the decision rules identified above when applied to lognormal distributions, which represent the idealized form assumed for environmental data in recent guidance (e.g. USEPA 1992, 1994). In the second paper of this series, decision rules are applied to data exhibiting characteristics that are more like those commonly observed among actual environmental data.

Methods

Because it is impossible to determine the "true conditions" that exist at a hazardous waste site, we based the evaluation of the decision rules presented in this paper on simulations. In these simulations, an underlying set of conditions is assumed, those conditions are then "sampled," summary statistics are derived from the "sample data," the required comparison is performed, and the results tabulated.

Definition of Simulations

All simulation programs were written in MicroSoft Fortran V5.1 and contain routines from the IMSL libraries. The IMSL routines were used for sampling from the distributions and for calculating the sample averages, sample standard deviations, and T-statistic based upper confidence limits and lower confidence limits. The Land-equation based upper and lower confidence limits were obtain using the BTNCTD program (Land et al. 1977). Other statistics (including the maximum measured value and the various percentiles) were programed directly.

The underlying distributions from which the simulated data were sampled in this study are lognormal distributions. Sets of 1, 5, 10, 20 and 100 samples were randomly selected from each distribution (generated for each of the means and CV's evaluated) for each iteration of each simulation. Test statistics were then generated from each sample set.

Means for the underlying lognormal distributions of 10, 20, 100, 200, 500, 1000, 2000, 5000, 10000, 20000, 100000, 200000 and 1000000 were evaluated in these simulations. The critical value was set at 1000; therefore, the ratios of the true means (i.e., the means of the underlying distributions) to the target value examined are 0.01, 0.02, 0.1, 0.2, 0.5, 1, 2, 5, 10, 20, 100, 200 and 1000. Each of the means was also paired with CV's of 0.5, 1.0, 2.0, 5.0, and 10.0 (representing the variability of the underlying distributions), which resulted in a set of 65 distributions that were sampled to evaluate each test statistic (and the associated decision rules) in this study.

Hypotheses and Statistical Tests

The set of hypotheses (null and alternate) evaluated using each of 9 different statistical tests was:

$$H_0: \ m < Q; \text{ and}$$
$$H_a: \ m \geq Q;$$

where m is the true mean of the distribution and Q is the target value.

The mathematical expressions for the test statistics used to evaluate these hypotheses are listed in Table 1. In every case, the test statistic (calculated as shown in the second column of Table 1) was compared to the target value and the null hypothesis was rejected if the test statistic was greater than the critical value. Otherwise the null hypothesis was accepted. For each set of conditions (i.e., a defined mean, CV, and sample size), at least 2,000 (and as many as 10,000) simulated data sets were produced (i.e., 2,000 to 10,000 iterations of each simulation were conducted) to evaluate performance.

Comparison of the Performance of the Various Hypothesis Tests

The tests are compared in terms of the "power curves" derived by evaluating the simulated data sets. That is, for each iteration of a simulation (with the underlying true conditions known and fixed), each test statistic was calculated and compared to the critical value. Results are reported as the number of iterations for which the null hypothesis is rejected divided by the total number of iterations. These proportions provide estimates of the probability of rejecting the null hypothesis for the stated set of conditions.

TABLE 1 -- Definitions for test statistics[a]

Test Statistic	Implementation
Sample Average	$AVG = \Sigma(x_i)/n$
T-Statistic Based Upper Confidence Limit[b]	$T\text{-}UCL_{95} = AVG + T(n\text{-}1,0.05)*\sigma_s/SQRT(n)$
T-Statistic Based Lower Confidence Limit[b]	$T\text{-}LCL_{05} = AVG - T(n\text{-}1, 0.05)*\sigma_s/SQRT(n)$
Simple Maximum Likelihood Estimate (Lognormal)[c]	$SMLE = exp(\mu_l + \sigma_l^2 / 2)$
Land-Equation Based Upper Confidence Limit[d]	$L\text{-}UCL_{95}$; Computed using BTNCTD program
Land-Equation Based Lower Confidence Limit[d]	$L\text{-}LCL_{05}$; Computed using BTNCTD program
Maximum Value	$MAX = x_n$
90th Percentile	x_n for samples sizes of 1 and 5, $(x_i + x_j)/2$ where i,j = (9,10), (18,19), (90,91) for samples sizes of 10, 20 and 100 respectively
95th Percentile	x_n for samples sizes of 1, 5 and 10, $(x_i + x_j)/2$ where i,j = (19,20), (95,96) for samples sizes of 20 and 100 respectively

[a] For the calculations shown above, consider the sampled data points to be sorted in ascending order, with "n" as the sample size (hence x_n is the maximum value). In addition,
$\sigma_s^2 = \Sigma(x_i - AVG)^2/(n\text{-}1)$
$\mu_l = \Sigma ln(x_i)/n$
$\sigma_l^2 = \Sigma(ln(x_i) - \mu_l)^2/n$
$T(n\text{-}1, 0.05) = $ 95th percentile of a T distribution with n-1 degrees of freedom.
[b] Values for t-statistics are available in any standard statistics text (see, for example, Bickel and Doksum 1977).
[c] Gilbert 1987
[d] Land et al. 1977

When the null hypothesis is true, acceptance of the null is equivalent to making the right decision; the rate at which this happens is the compliment of the "Type I" (false-positive) error rate (i.e. the rate at which the null hypothesis is falsely rejected). When the null hypothesis is *not* true, acceptance of the null hypothesis constitutes a Type II error and the rate at which such false-negative errors occur is the compliment of the power of the test (i.e. the rate at which the null hypothesis is correctly rejected) for the conditions under consideration. The tests are compared via graphs that represent the probability of rejecting the null hypothesis as a function of the underlying true conditions, in particular: the true mean relative to the critical value. A "good" test is one that exhibits a high rate of rejection when the null hypothesis is not true, and exhibits a low rate of rejection when the null hypothesis is true.

For each of the tests under consideration, "power" curves are approximated by interpolating straight lines between the estimates of the probabilities of rejection obtained at the various values of the means that were actually examined (which are listed above).

Results and Discussion

The results of the simulations described above are presented as a series of power curves. Among other things, the curves presented demonstrate that the degree of statistical control achieved by a particular decision rule is independent of the choice of the form of the null hypothesis. The loss of statistical control that occurs when a particular decision rule is applied to a data set that does not satisfy the rule's validity criteria is also clearly demonstrated. Some of the rules evaluated do not offer the formal statistical control that is an inherent design feature of other rules. Nevertheless, results indicate that such "informal" decision rules may provide superior overall control of error rates, when their application is restricted to data exhibiting particular characteristics.

Hypothetical Power Curve

A power curve depicts the relationship between the probability of rejecting the null hypothesis and the true condition (i.e. mean and CV) for a given decision rule. To illustrate, a hypothetical curve is presented in Figure 1. The X-axis in Figure 1 indicates one dimension of the true condition (expressed in terms of the ratio of the true mean, m, to the target value, Q). Note that it is presented on a logarithmic scale. The Y-axis is the probability of rejecting the null hypothesis (i.e. that $m < Q$). A "perfect" power curve would indicate zero probability of rejecting the null hypothesis whenever the ratio of m to Q is less than one and the probability would rise to one for all values of this ratio greater than one. This "perfect" (errorless) power curve is depicted as a dotted line in Figure 1.

A more realistic power curve is depicted as the solid line in Figure 1. By comparing the solid line to the dotted line, the effects of error become apparent. To the left of one on the X-axis, the vertical distance between the solid curve and the X-axis (which is also the line representing zero probability of rejecting the null) represents the rate at which a false positive (Type I) error can occur. Thus, for this hypothetical curve, when m is one tenth of Q (i.e. at 0.1 on the X-axis), the solid curve indicates an 8% chance of *falsely* rejecting the null. This error increases as the ratio m/Q approaches one.

Once m/Q exceeds one in Figure 1, the null hypothesis should be rejected. Therefore, to the right of one on the X-axis, the vertical distance between the solid line

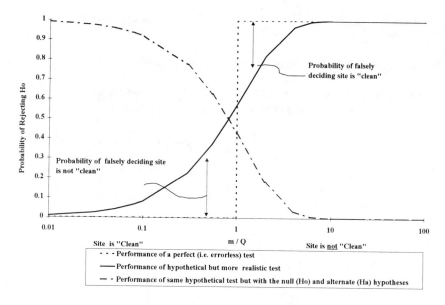

Figure 1: Power Curve for Hypothetical Test with Ho: m < Q and Ha: m ≥ Q

and the line representing unit probability of rejecting the null represents the false negative (Type II) error rate. As depicted, this error rate is vanishingly small when the m is 10 times Q but increases steadily as m approaches Q from the right. The power curves presented in the remaining figures, which depict the performance of the decision rules evaluated in this paper, can be interpreted in precisely the same manner as Figure 1.

One additional feature of Figure 1 deserves mention. The dashed line in the figure represents a power curve for a hypothetical decision rule that is identical to that assumed for the solid line except that the dashed curve represents testing a reversed set of hypotheses (i.e. the alternate hypothesis for the solid line that m > Q is taken as the null for the dashed line with the old null serving as the new alternate hypothesis). This is presented to illustrate that reversing hypotheses serves precisely to transform the power curve by reflection across the line defined by 50% probability of rejecting the null. This relationship holds because the two (complimentary) decision rules would be tested using the same data set for the same true condition so that the value of the summary statistic would be the same. Thus, all that changes is that values formerly indicating rejection of the null now indicate acceptance of the null and vice versa.

The relationship between a power curve for a particular decision rule and one for the same decision rule with the hypotheses reversed was confirmed during the simulations conducted in this study. Because of this relationship, it is not necessary to present an independent set of power curves for decision rules involving reversed hypotheses; these can simply be generated by the indicated reflection transformation. One must simply adjust for the nature of the error defined. Thus, for example, a significance level of 5% for one null translates to a significance level of 95% (100% - 5%) for the case with the hypotheses reversed.

The Performance of Decision Rules Applied to Lognormally Distributed Data

The performance of tests to determine whether m < Q that employ comparisons between direct, best estimates of m and Q are depicted in Figures 2a and 2b.

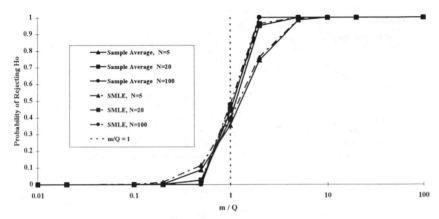

Figure 2a: Performance of Tests of Ho: m < Q that Involve Comparisons of Direct (Unbiased) Estimates of the Mean (CV = 2)

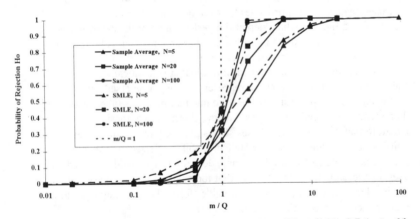

Figure 2b: Performance of Tests of Ho: m < Q that Involve Comparisons of Direct (Unbiased) Estimates of the Mean (CV = 5)

In Figures 2a and 2b, solid lines represent power curves for the common arithmetic average (AVG), which is calculated in the traditional manner (Table 1). This estimator of the sample mean is unbiased for any type of data distribution. Dashed lines indicate the performance of the simplified maximum likelihood estimator (SMLE) for a lognormal distribution (Table 1), which has been recommended for use when data are assumed to exhibit lognormal behavior (e.g. USEPA 1992, 1994). Note that the SMLE is known to be a biased estimator of the mean (Attfield and Hewett 1992). However, the bias is small and often ignored.

Comparing the dashed and solid curves in Figure 2a, for example, it is apparent that the SMLE and the AVG are comparable even for data sets containing as few as five measurements (when the CV of the underlying data is as small as 2). As the dispersion (variability) of the data increases (i.e. as the CV increases), however, differences between the SMLE and the sample average (for the smallest data sets) become more apparent. This is indicated by the greater differences between corresponding dashed and solid curves in Figure 2b then Figure 2a.

The curves depicted in Figures 2a and 2b indicate that employing either the AVG or the SMLE in tests of the mean provide balanced false-positive and false-negative error rates (i.e. the curves are relatively symmetric about the vertical line representing the point at which m just equals Q. Further, as long as data exhibit a CV of 2 or less (Figure 2a), as few as 20 samples appear to provide sufficient power to assure less than a 5% chance either of falsely concluding that a contaminated site is clean or of falsely concluding that a clean site is contaminated when m varies from Q by a factor of two or more. Even for sample sets as small as five, both false-positive and false-negative error rates as small as 5% are achieved when the true mean varies from the target by at least a factor of five. As is commonly the case, more samples are required to maintain similar error rates when the data exhibit greater variability (Figure 2b).

Because the AVG and the SMLE are not designed to provide formal statistical control of error (i.e. the significance levels for these tests are not fixed by the design of the test), decision rules involving best estimates of the mean have not been considered adequately protective of public health. Consequently, tests employing upper or lower confidence limits to the mean (UCL's or LCL's, respectively) have been preferred.

Figures 3a and 3b present a series of power curves for decision rules incorporating

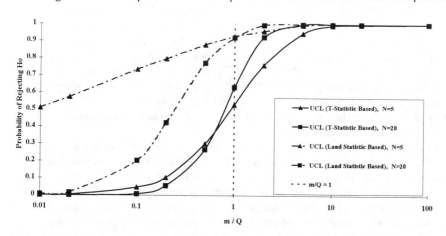

Figure 3a: Peformance Tests of Ho: m< Q that Involve Comparisons with Biased Estimates of the Mean for CV = 5 (UCL Tests)

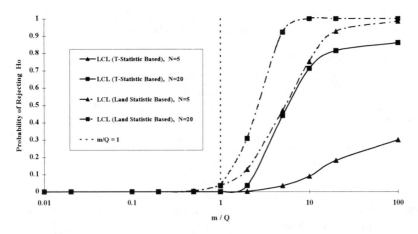

Figure 3b: Peformance Tests of Ho: m< Q that Involve Comparisons with Biased Estimates of the Mean for CV = 5 (LCL Tests)

either UCL's or LCL's. The solid lines in this figure represent 95% UCL's and LCL's derived using the Student's t-statistic (T-UCL$_{95}$ and T-LCL$_{05}$, respectively, see Table 1) and the dashed lines represent 95% UCL's and LCL's derived using the Land Equation (L-UCL$_{95}$ and L-LCL$_{05}$, respectively, see Table 1).

Among other things, the curves presented in these figures provide a striking illustration of the importance of employing statistical procedures that are appropriate to the characteristics of the data being evaluated. Because the data sets evaluated in this study were drawn from an underlying lognormal distribution, tests employing the L-UCL$_{95}$ and L-LCL$_{05}$ exhibit the expected statistical control. Tests employing the T-UCL$_{95}$ and L-LCL$_{05}$ do not.

Formally, a 95% UCL represents an estimate of the mean that should be larger than the true mean 95% of the time (i.e. 95% of the UCL's estimated from random sample sets collected from an underlying distribution will be larger than the true mean of the underlying distribution, assuming that the UCL is calculated appropriately). Therefore, when m just equals Q (i.e. at one on the X-axis of Figure 3a), tests involving the 95% UCL should reject the null precisely 95% of the time. This is at least approximately true for L-UCL$_{95}$ but is clearly false for T-UCL$_{95}$ in Figure 3a. As a consequence, regular application of goodness-of-fit tests to determine the form of the sampled distribution have been strongly recommended (Berman 1995).

Similar behavior is observed among the 95% LCL's in Figure 3b. Under statistical control, 95% LCL's should exhibit a significance of 5% (100% - 95%). As indicated in the figure, the L-LCL$_{05}$ provides statistical control, the T-LCL$_{05}$ does not. The curves in Figures 3a and 3b also confirm that the statistical control provided by the L-UCL$_{95}$ and L-LCL$_{05}$ is independent of sample size. Thus, for example, curves representing application of the L-UCL$_{95}$ for a five sample set and a 20 sample set, respectively, both pass through the point representing 95% probability of rejecting the null at the line representing one on the X-axis.

It is precisely because of the statistical control provided for error related to health protectiveness that the L-UCL$_{95}$ has been recommended in recent guidance (e.g. USEPA 1992, 1994).   However, the tradeoff for using this decision rule is also apparent from Figure 3a. False-positive error rates associated with this decision rule are large. In fact, if data exhibit a CV of 5 (Figure 3b) and only five samples are collected to support the analysis, a site will be falsely determined to require cleanup approximately 50% of the time when m at the site is as little as 1% of Q. Even if 20 samples are employed to support the analysis, a site at which m is one tenth of Q will be falsely determined to be contaminated 20% of the time.

Figures 4a and 4b present power curves for decision rules incorporating defined percentiles of the sampled distribution and those incorporating the maximum measured

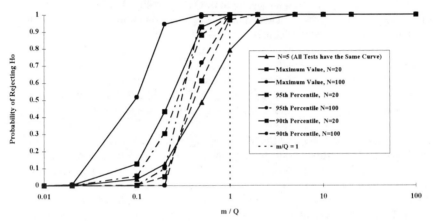

Figure 4a: Performance of Tests of Ho: m < Q that Involve Percentiles of the Sample Population (CV = 2)

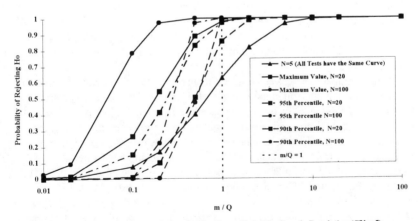

Figure 4b: Performance of Tests of Ho: m < Q that Involve Percentiles of the Sample Population (CV = 5)

value (MAX). The solid lines in this figures represent curves for the MAX. The dashed lines represent curves for the 95th percentile (%95) and the lines composed of both dashes and dots represent curves for the 90th percentile (%90). Values for these summary statistics are determined as described in Table 1.

The curves in Figured 4a and 4b demonstrate that the MAX behaves differently from most of the other tests evaluated in this study. Unlike other tests (including tests involving percentiles), the false-positive error rate actually *increases* with increasing sample size for the MAX. This is expected because, the larger the number of samples collected the greater the chance that one extremely large value will be encountered that would require that the null hypothesis be rejected. Thus, if a 100 samples are drawn from a distribution exhibiting a CV of five, there is an 80% chance of falsely concluding that m is larger than Q when m is no more than one tenth of Q (Figure 4b). In contrast, the chance of reaching this same false conclusion under the same conditions falls to 25% if the MAX test is performed on a data set containing only 20 samples.

Figures 4a and 4b also indicate that decision rules employing %90 or %95 share one very important characteristics with tests employing UCL's; they tend to better control false negative errors than false positive errors. In fact, even for data exhibiting a CV as large as five, for a sample size of 20, tests involving the %90 and %95 all prevent greater than a 3% chance of falsely determining a site to be clean when the true mean is twice the target value (Figure 4b).

Intuitively, the %90, %95, and the MAX test should all be biased toward minimizing false-negative errors relative to tests involving best estimates of the mean because, for most distributions, the observed values of such statistics should be larger than the true mean with high probability. For positively skewed distributions such as lognormal distributions, however, m shifts to values representing higher percentiles of the distribution as the spread (CV) of the distribution increases. For example, calculations indicate that for lognormal distributions exhibiting a CV of 10, m assumes a value equivalent to the 86th percentile of the distribution. For distributions exhibiting smaller spreads, m assumes values closer to the median (i.e. the 50th percentile). Therefore, over the range of distributions most likely to be relevant to environmental data, %90 and %95 tests will be biased to minimize falsely determining that a site is clean. Clearly, the MAX test will always be biased in this manner.

Based on a comparison between Figures 3a and 4b, one might conclude that percentile tests offer a potential advantage over tests involving UCL's. For example, tests based on data sets containing 20 or more samples and employing, respectively, the %90 and the L-UCL$_{95}$ control false-negative errors equally well. However, comparing the false-positive error rates for these tests when the true mean is one tenth of the target value indicates a 20% chance of falsely concluding that a site is contaminated when the L-UCL$_{95}$ is employed (Figure 3a) but only a 2% chance of reaching the same false conclusion if the %90 is employed (Figure 4b). However, percentile tests do not provide the formal statistical control of error that UCL tests provide (*as long as the UCL is derived in a manner that is appropriate to the underlying distribution of the data*). Thus, like tests incorporating best estimates of the mean, percentile tests have traditionally been limited to screening type analyses.

Conclusions

The results of this study demonstrate that the performance of the decision rules applied to environmental data can be evaluated and compared so that procedures that best minimize potential errors can be identified. All of the results reported in this study, however, are limited to an evaluation of data sets that are lognormally distributed, while data derived from actual environmental studies may not generally be as well behaved. As indicated previously, for example, environmental data may include a mix of zero concentrations and non-zero concentrations and such data are frequently censored. There is also no reason that detectable environmental concentrations be constrained to behave as if they were lognormally distributed.

As clearly demonstrated in this study, formal statistical control is realized *only* when data satisfy the validity criteria for the decision rules applied. Therefore, before the decision rules that are applied to environmental data can be optimized, it will be necessary to evaluate their behavior over a range of characteristics that bracket the characteristics commonly exhibited by field data. Only then will it be possible to demonstrate that the statistical control associated with any particular decision rule can be maintained over data exhibiting a sufficient range of characteristics that such a decision rule can be considered robust and applied broadly. Barring that, compromises may be required: a set of decision rules may exist such that each member of the set may be shown to be valid under a subset of the conditions of interest. It may also be necessary to accept less than formal statistical control as long as particular decision rules can be demonstrated to be reasonably well-behaved over a well defined set of conditions that are applicable to a range of types of environmental studies.

The behavior of candidate decision rules as they are applied to data sets exhibiting a range of characteristics that more closely mimic environmental data than the data sets investigated in this study is the subject of the second paper of this series (Berman et al. 1998).

References

Attfield, M.D. and Hewett, P., 1992, "Exact Expressions for the Bias and Variance of Estimators of the Mean of a Lognormal Distribution," *Amer Ind Hyg Assoc J* 53:432-435.

Berman, D.W., 1995, "Does Risk Assessment Work? Limitations Imposed on risk Assessment by Data Quality and Common Practice," *Challenges and Innovations in the Management of Hazardous Waste, Proceedings of an Air &Waste Management Association and Waste Policy Institute Conference, Pittsburgh, Pennsylvania,* pp. 493-503.

Berman, D.W., Allen, B.C. and Van Landingham, C.B., 1998, "Risk Assessment and Risk Management Under Superfund. Part 2: How Often Do We Really Do the Right Thing?" *Third Symposium on Superfund Risk Assessment in Soil Contamination*

Studies, ASTM STP 1338 K.B. Hoddinott, Ed., American Society for Testing and Materials.

Bickel, P.J. and Doksum, K.A., 1977, *Mathematical Statistics: Basic Ideas and Selected Topics*, Holden-Day, Inc., San Francisco.

Box, G.E.P., Hunter, W.G., and Hunter, J.S., 1977, *Statistics for Experimenters: An Introduction to Design, Data analysis, and Model Building*, John Wiley and Sons, New York.

Gilbert, R.O. 1987, *Statistical Methods for Environmental Pollution Monitoring*, Van Nostrand Reinhold, New York.

Land C.E., 1971, "Confidence Intervals for Linear Functions of the Normal Mean and Variance," *Annals of Mathematical Statistics* 42:1187-1205.

Land C.E., Greenberg L.M., Hall C., Drzyzgula C.C, 1977,. *BTNCTD: Exact Confidence Limits for Arbitrary Linear Functions of the Normal Mean and Variance*, Radiation Epidemiology Branch, National Cancer Institute and Information Management Systems

The U.S. Environmental Protection Agency, May 1992, *Supplemental Guidance to RAGS: Calculating the Concentration Term*, Office of Solid Waste and Emergency Response, Publication 9285.7-081.

The U.S. Environmental Protection Agency, December 1994, *Draft Soil Screening Guidance* Vol. 59, No. 250 (59FR67706).

D. Wayne Berman[1], Bruce C. Allen[2], and Cynthia B. Van Landingham[3]

EVALUATION OF THE PERFORMANCE OF STATISTICAL TESTS USED IN MAKING CLEANUP DECISIONS AT SUPERFUND SITES. PART 2: REAL WORLD IMPLICATIONS OF USING VARIOUS DECISION RULES

REFERENCE: Berman, D.W., Allen, B.C., and Van Landingham, C., **"Evaluation of the Performance of Statistical Tests Used in Making Cleanup Decisions at Superfund Sites. Part 2: Real World Implications of Using Various Decision Rules"** *Superfund Risk Assessment in Soil Contamination Studies: Third Volume, ASTM STP 1338*, K.B. Hoddinott, Ed., American Society for Testing and Materials, 1998.

ABSTRACT: This is the second paper of a study in which decision rules commonly employed to determine the need for cleanup are evaluated both to identify the conditions under which they may lead to erroneous conclusions and to quantify the rate at which such errors may occur. In this paper, performance is evaluated for such rules when applied to data exhibiting a range of characteristics commonly exhibited by environmental data. Results are reported for simulations involving data exhibiting normal distributions, lognormal distributions, and P-lognormal distributions (lognormal distributions with additional, finite density at zero). Some of the data sets employed were also censored.

Results indicate that none of the decision rules commonly applied to environmental data provide the advertised statistical control over the complete range of characteristics commonly exhibited by field data from hazardous waste sites. However, rules derived for normally distributed data or (modified from those derived for such data) appear to provide reasonable control of error rates that, appropriately, improve with increasing sample size. Also, because the advertised statistical control is seldom realized, it does not appear valid to prefer tests for which such formal control is inherent. Decision errors may best be controlled by matching selected decision rules to the observed characteristics of data.

KEYWORDS: risk assessment, site assessment, statistics, decision error, error rates, hypothesis tests, confidence limits

[1]President, Aeolus Environmental Services, 751 Taft St., Albany, CA 94706.

[2]Project Manager, ICF Kaiser Engineers, P.O. Box 14348, Research Triangle Park, NC, 27709.

[3]Project Manager, ICF Kaiser Engineers, 602 East Georgia, Ruston, LA 71270.

This is the second of two papers in which decision rules commonly employed to determine the need for cleanup at Superfund sites are evaluated both to identify conditions under which they may lead to erroneous conclusions and to quantify the rates at which such errors may occur. In the first paper (Berman et al. 1998), considerations that need to be addressed to evaluate performance are introduced and performance is judged for cases in which the procedures are applied to data exhibiting reasonably ideal behavior (i.e., the data are lognormally distributed).

In this paper, performance is evaluated for cases in which the various decision rules are applied to data exhibiting a range of characteristics that more closely mimic what is observed in environmental studies. .

Background

As indicated in the first paper (Berman et al. 1998), the decision whether to remediate a hazardous waste site typically reduces to determining whether the arithmetic mean of some contaminant concentration exceeds a critical target value. The decision rule typically incorporates a hypothesis test in which it is assumed as a null hypothesis that the true mean concentration, "m," is less than the critical target, "Q" (i.e. $m < Q$) and this is evaluated by comparing Q to some estimate of m (which is derived from field measurements) to determine whether the null hypothesis should be rejected in favor of an alternate hypothesis that $m \geq Q$.

Estimates of m that are or have been used to determine the need for cleanup (and that are evaluated in this paper) include:

- the common arithmetic average (AVG);
- the simplified maximum likelihood estimator (SMLE) for the mean of a lognormal distribution (Gilbert 1987);
- the Land-Equation (Land 1971) derived 95% upper confidence limit and lower confidence limit for the mean of a lognormal distribution (L-UCL$_{95}$ and L-LCL$_{05}$, respectively);
- the t-statistic (Bickel and Doksum 1977) derived 95% upper confidence limit and lower confidence limit for the mean of a normal distribution (T-UCL$_{95}$ and T-LCL$_{05}$, respectively); and
- the Chen-Method (Chen 1995) derived estimates of the 95%, 90%, and 80% lower confidence limit for the mean of a positively skewed distribution (Chen$_{05}$, Chen$_{10}$, and Chen$_{20}$, respectively).

While not representing formal estimates of the mean of a distribution, the maximum measured value (MAX) and the 90th and 95th percentiles of the measured values (%90 and %95, respectively) are also evaluated because they have been used as screening statistics for hazardous waste sites.

Decision rules employing the Chen Method have been added to this discussion because they have been recommended in recent guidance (USEPA 1996). Such rules were not discussed in the first paper.

The performance of each of the decision rules evaluated in this study is described in terms of the rule's error rates (i.e. the probabilities that the rule leads to incorrect decisions given a defined set of conditions). The degree of control provided for both Type I and Type II errors is addressed.

For the formulation of the null hypothesis evaluated, Type I errors would result in a false determination that a site is contaminated with the associated consequence that money and resources may be wasted in cleaning up a site that is acceptably clean. A Type II error would result in a false determination that a site is clean, which may have adverse health consequences. The choice of null and alternate hypotheses can also be reversed in a decision rule, which would reverse the definitions of Type I and Type II errors. As shown in the first paper, however, this has no effect on the degree or type of statistical control provided by a decision rule, other than the effect of changing terminology. Thus, only the above stated formulation of the null and alternate hypotheses is addressed in this paper.

Also as indicated in the first paper, several of the statistical tests evaluated as decision rules in this study offer formal statistical control of one of the two types of error *when applied under conditions that satisfy the validity criteria for the test.* Formal statistical control means that the null hypothesis is rejected at a fixed and defined rate at the decision point (i.e. when m = Q) and this rejection rate is independent of sample size or other conditions. The rejection rate at the decision point is defined as the significance or α (alpha) level of the test.

The type of error that is not controlled by the significance level of a test, is typically controlled by the power of the test. Power is the probability of correctly rejecting the null hypothesis for any defined value of m (other than m = Q). Importantly, power is dependent both on the condition under which the test is conducted (i.e. the value of m) *and* on the number of samples in the data set to which the test is applied. Note that, for the decision rules evaluated in this study that do not provide *formal* statistical control, control of both types of error depend on power. The power of a well-behaved test increases with increasing sample size.

Because characteristics of environmental data frequently fall at the fringe of acceptability for the validity criteria of the tests evaluated in this study, the "advertised" statistical control for particular tests may not be realized. Therefore, the error rates actually achieved when such tests are applied to real environmental data are evaluated in this paper and recommendations are provided for minimizing the resulting decision errors.

Methods

As indicated in the first paper, the evaluation of decision rules presented in this paper is based on simulations. Because it is impossible to determine the "true conditions" that exist at a hazardous waste site, the evaluation of the decision rules presented in this paper had to be based on simulations. In these simulations, an underlying set of conditions is assumed, those conditions are then "sampled," summary statistics are derived from the "sample data," the required comparison is performed, and the results tabulated.

Definition of Simulations

The simulations used to support this paper were conducted as described in the first paper (Berman et al. 1998). For this study, however, the underlying distributions from which the simulated data were sampled include P-lognormal distributions (i.e. lognormal distributions with either 30% or 70% of the density of the distribution added at zero) in addition to the lognormal distributions addressed in the first paper. Sets of 1, 5, 10, 20, and 100 samples were randomly selected from each distribution (generated for each of the means and coefficients of variation, CV's, evaluated) for each iteration of each simulation. Test statistics were then generated from each sample set.

For most of the cases in which test statistics were evaluated using data sampled from P-lognormal distributions, 30% of the density of the P-lognormal distribution was added at zero and means for the underlying distributions of 100, 500, 1000, 2000, and 5000 were included in the simulations. As in the first paper, the critical value ,Q, was set equal to 1000. Each of these means was also paired with CV's of 0.5 (for some cases), 1 (for some cases), 2, 5, 10, and 20 (for some cases). This resulted in a set of an additional 25 distributions that were sampled to evaluate each test statistic (and the associated decision rules). In a select and very limited number of cases, a few of these simulations were also repeated with data sampled from an underlying P-lognormal distribution with 70% density at zero.

A subset of the above simulations were repeated with the data sets modified by censoring. Both lognormal distributions (from the original study) and P-lognormal (with 30% density at zero) distributions were sampled and censored for these cases. Censoring was accomplished by setting a censoring point and replacing all samples returning values less than the censoring point with a new value set equal to one-half of the censoring point. Censoring points equal to 0.01%, 0.1%, 1%, and 10% of Q (1000) were evaluated. Means for the underlying distributions that were examined in these cases include: 10, 100, 500, 1000, 2000, and 10000 for lognormally distributed data and 100, 500, 1000, 2000, and 5000 for P-lognormally distributed data. These were each paired with CV's of 0.5 (for some cases), 1 (for some cases), 2, 5, 10, and 20 (for some cases). This resulted in a set of an additional 25 distributions that were sampled and censored in each of four ways to evaluate each test statistic (and the associated decision rules).

Hypotheses and Statistical Tests

As in the first paper, the set of hypotheses (null and alternate) evaluated based on each of 12 different statistical tests is:

$$H_0: \ m < Q; \text{ and}$$
$$H_a: \ m \geq Q;$$

where m is the true mean of the distribution and Q is the target critical value.

Mathematical expressions for all of the statistical tests used to test these hypotheses (except those incorporating Chen Method statistics) are listed in Table 1 of the

first study. The test statistics (Chen$_\alpha$) proposed by Chen (1995) to account for skewness
are derived as modifications to the t-statistic (Bickel and Doksum 1977).
    As implemented by the U.S. Environmental Protection Agency (1996), the Chen
statistic is:

$$t_2 = t + a(1 + 2t^2) + 4a^2(t + 2t^3),$$

$$a = (n^{\frac{1}{2}}/6)*\Sigma(x_i - \bar{x})^3/[(n-1)(n-2)s^3];\ \text{and}$$

$$s^2 = \Sigma(x_i - \bar{x})^2/(n-1);$$

where:
    t   is the usual Student's t-statistic;
    a   is a function of the skewness of the data and is calculated as indicated;
    s   is the sample standard deviation; and
    other symbols are defined in Table 1 of the first paper.

    For the Chen test, the null hypothesis that m < Q is rejected in favor of the
alternate when $t_2 > z_\alpha$, where z_α is the $100(1 - \alpha)$th percentile of a standard normal
distribution.
    The test statistics for the decision rules evaluated are calculated either as shown
above or as shown in Table 1 of the first paper. For those shown in Table 1, the
calculated test statistic was compared to Q and the null hypothesis was rejected if the test
statistic was greater than Q. Otherwise the null hypothesis was accepted. For each set of
conditions (i.e. a defined mean, CV, and sample size), at least 2,000 (and as many as
10,000) simulated data sets were produced (i.e., 2,000 to 10,000 iterations of each
simulation were conducted) to evaluate performance.

Comparison of the Performance of the Tests

    The performance of each test is evaluated as described in the first paper. Results
are reported as the number of iterations for which the null hypothesis is rejected divided by
the total number of iterations. These proportions provide estimates of the probability of
rejecting the null hypothesis for the stated set of conditions.
    In the first paper, the performance of each decision rule is presented in the form of
a power curve. Because the goal in this paper is to evaluate performance over a range of
data sets with varying characteristics, bar charts are presented in which performance at a
single, specified true condition (i.e., a specified true mean) is presented simultaneously for
several decision rules applied to data sets exhibiting a range of characteristics.

Results and Discussion

Results of the simulations conducted to evaluate performance are presented in a series of
bar charts in this section. Each chart indicates the extent that the statistical control

achieved by a particular decision rule varies as a function of the characteristics of the data set being evaluated.

Decision Rules Based on Best Estimates or Upper Bound Confidence Limits for the Mean

Figure 1 indicates the probability of rejecting the null hypothesis that m < Q when m just equals Q (i.e. the significance level that is achieved for each of the decision rules applied to each of the data sets presented). The Y-axis in Figure 1 indicates the numerical probability of rejecting the null hypothesis. Each of the bars in the figure represent a particular decision rule applied to a set of data that exhibit the specific characteristics defined on the X-axis (i.e. the number of measurements in the data set and the true CV of the data set). The data sets depicted in this figure were all derived from sampling an ideal lognormal distribution (with m = Q and the CV indicated on the X-axis).

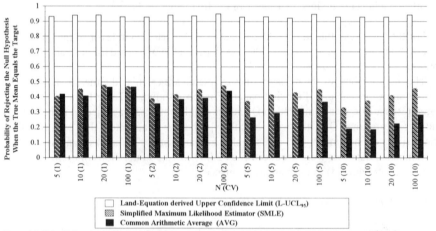

Figure 1: Alpha Values for the Indicated Decision Rules Applied to Data Sets Sampled From a Lognormal Distribution

The white bars in Figure 1 represent the probability of rejecting the null hypothesis for each of the indicated data sets using a decision rule that incorporates a comparison between Q and an estimate of m represented by the L-UCL$_{95}$. The cross-hatched bars indicate the probability of rejecting the null using a decision rule incorporating a comparison between Q and an estimate of m derived using the SMLE and the black bars indicate the probability of rejecting the null with a decision rule incorporating the AVG. Note that the L-UCL$_{95}$, the SMLE, the AVG, and all of the other test statistics discussed in this section are defined in the Background Section of this paper.

Several interesting observations and conclusions are immediately apparent from Figure 1. First, the frequency with which the null hypothesis is rejected using the L-UCL$_{95}$ is very high and remains very close to 95% independent of sample size or the spread of the data (at least up to CV's as large as 10). This simply confirms the type of statistical control for which the L-UCL$_{95}$ is designed (Gilbert 1987).

The L-UCL$_{95}$ is an estimate of the mean of a lognormal distribution that is intended to be larger than the true mean 95% of the time. Thus, the null hypothesis (m < Q) is rejected using this estimator 95% of the time when the true mean is just equal to Q. When the L-UCL$_{95}$ is employed in the decision rule and data are truly lognormally distributed, there should never be greater than a 5% (100%-95%) chance of falsely deciding that the mean concentration is less than Q. However, whether one defines this as control of Type I or Type II error depends on the choice of the formulation of the null and alternate hypotheses.

The rejection rates reported in Figure 1 for the L-UCL$_{95}$ actually vary somewhat but appear consistently to be slightly less than 95% for the conditions reported. No consistent trend is observed across sample size or across data spread (CV). This suggests that, even when the L-UCL$_{95}$ is applied to lognormally distributed data, the α level that is achieved is somewhat less than that advertised for this statistic. A similar effect is also observed among results for the L-LCL$_{05}$ statistic. If these observations are valid, however, the deviations are small (i.e. no more than a couple of percent) and there is no obvious explanation because the Land Equation is supposed to provide exact estimates of confidence limits for lognormally distributed data (Land 1971).

Most of the variability among the α levels observed for the L-UCL$_{95}$ (and the L-LCL$_{05}$) is believed attributable to the limited stability of the simulations performed because of the relatively small number of iterations (2,000) employed when evaluating the L-UCL$_{95}$ and L-LCL$_{05}$ tests. Fortunately, the speed of the calculations allowed simulations involving other test statistics to incorporate 10,000 iterations so that such variability is substantially reduced for the other cases examined.

In contrast to the L-UCL$_{95}$, neither the AVG estimates nor the SMLE estimates of the mean are designed to provide formal statistical control of error rates. Thus, as seen in Figure 1, the α levels for these tests are much more highly variable than the α levels exhibited by the L-UCL$_{95}$. Because these two estimators are supposed to provide best estimates of m and m in the figure is set equal to Q, one might expect that the rejection rate for these tests should approach 50% (i.e. best estimates of the mean should be larger than true mean about half of the time). As clearly depicted in the figure, however, this is not the case.

This effect is due to the asymmetry of a lognormal distribution. Because the median is less than the mean, when sampling such a distribution, the probability of obtaining any single measurement below the mean is greater than obtaining any single measurement that is above the mean. Thus, the number of times that the average of a small data set will be less than the mean is somewhat greater than the number of times that the average of a small data set will be greater than the mean. Consequently the rejection rate observed in Figure 1 is less than 50%. In contrast, the *magnitude* of the means estimated for each data set may also be raised with infrequent, very high sampled values mixed in with more frequent, lower values. Therefore, the mean of such means should approach the true mean.

Actually, the SMLE is known to be a slightly biased estimator of the mean of a lognormal distribution (Attfield and Hewett 1992). It is in fact a computationally simplified adaptation of the exact maximum likelihood estimator for the mean of a lognormal distribution (Gilbert 1987). However, it has been shown that the extent of the bias is small (less than a few percent) for the lognormal data sets exhibiting the range of characteristics examined in this study. As indicated previously, this range of characteristics is also likely to bracket the vast majority of the kinds of data sets derived from field measurements. Thus the exact maximum likelihood estimator has seldom been applied to environmental data.

The effect caused by the asymmetry of lognormal distributions attenuates as sample sizes are increased. Collecting a larger number of samples from which to estimate the mean of the distribution increases the probability that at least some high values will be included with the expected larger number of lower values. Thus, the rejection rate approaches 50% as decision rules involving both the AVG and the SMLE estimates of the mean are applied to sample sets containing increasing numbers of samples.

It is also immediately apparent from Figure 1 that the α level for the AVG is always smaller (i.e. further from 50%) than that observed for the SMLE. This is because, while the AVG is an unbiased estimator of the mean of *any* distribution (Box et al. 1978), the SMLE is derived specifically for lognormal distributions and is expected to provide a closer estimate for lognormal distributions than the AVG. Not surprisingly, differences between these two estimators are small for data sets exhibiting narrow spreads (i.e. with CV's of two and smaller). This is because the characteristics of such distributions are much closer to normal distributions, for which the AVG is a maximum likelihood estimator (Bickel and Doksum 1977), than lognormal distributions exhibiting larger spreads (i.e. with CV's of five or greater).

Figure 2A and 2B are presented to indicate the effects that distorting a lognormal

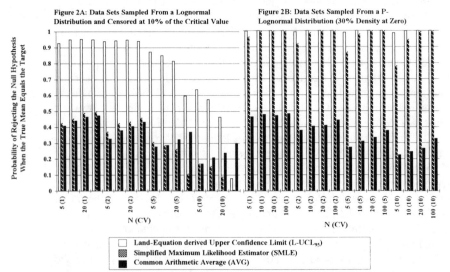

Figure 2: Alpha Values for the Indicated Decision Rules Applied to Data Sets Sampled From the Indicated Distributions

distribution and censoring data have on the error rates achieved by decision rules incorporating the L-UCL$_{95}$, AVG, or SMLE statistics. Each of the bars in these figures represent the same combination of decision rules, sample sizes, and CV's as those depicted in Figure 1. However, the methods used to derive the data sets differ from the methods employed for Figure 1. In Figure 2A, the underlying distributions are still lognormal distributions but the data sets are assumed to be censored so that any of the sampled values that are less than 0.1Q (i.e. 10% of the target) are assumed to be non-detected values and were replaced by a value of 0.05Q (i.e. one-half of the "detection limit").

Figure 2B is derived by sampling P-lognormal distributions, which are lognormal distributions with finite density added at zero. For this figure, 30% of the density of the entire distribution is assumed to exist at zero. This is equivalent to assuming that a field is representatively sampled and that 30% of the area sampled is *un*contaminated so that samples collected from that part of the field yield concentrations that are equal precisely to zero.

Results depicted in Figures 2A and 2B demonstrate clearly that the statistical control intended by use of the L-UCL$_{95}$ fails when this summary statistic is derived from data that are *not* adequately described by lognormal distributions. In Figure 2A, such effects are caused by censoring the data and this results in a decrease in α levels as the spread (CV) of the data increases. Importantly, the effect gets larger as sample size is increased, which is contrary to the general rule that larger sample sizes can be used to improve precision and reduce errors.

Note that, due to the manner in which data sets are censored in this study, the fraction of values actually censored in the data sets examined varies as a direct function of the spread of the data. For example, a censoring point that is set at 10% of the true mean of a lognormal distribution exhibiting a CV of 1 is approximately equivalent to the first percentile of the distribution. It is therefore expected that only approximately 1% of the samples from data sets exhibiting this spread will be censored. In contrast, the value equal to 10% of the true mean of a lognormal distribution exhibiting a CV of 2, 5, or 10 represents, respectively, the 12th, 35th, and 60th percentiles of those distributions. Correspondingly, it is expected that approximately 12%, 35%, and 60% of the samples derived will be censored when the underlying distributions have CV's of 2, 5, and 10, respectively. Based on an overview of the magnitude of effects observed for the complete range of combinations of censoring points and CV's examined in this study, it appears that effects due to censoring become important once 10 to 15% of the samples are censored.

Figure 2B indicates that the effect on the L-UCL$_{95}$ statistic caused when a lognormal distribution is distorted by adding density at zero is opposite to that observed in association with censoring. In this case, α levels increase to levels close to 100% from the intended 95%. This effect is observed immediately and strikingly for data sets exhibiting all sample sizes (i.e. 5 to 100) and all spreads (i.e. CV's between 1 and 10). As with the effects of censoring, this effect also increases (i.e. gets worse) with sample size, which again violates the general rule that larger sample sizes can be used to improve precision and reduce errors.

Although not depicted, effects observed when censoring a P-lognormal distribution appear to represent a combination of the effects observed either when a lognormal is distorted by adding density at zero or with censoring alone. Because of the way that censoring was applied in this study, the effect due to distorting the underlying distribution dominates for data sets with smaller spreads (i.e. with CV's of 2 and smaller) where the fraction of the non-zero samples that are censored is small (i.e. less than 10%). Obviously, 100% of the zero value samples are censored. The effect due to censoring dominates for data sets with larger spreads (i.e. with CV's of 5 and greater) where the fraction of the non-zero samples that are censored is larger. In the former cases, the α levels are observed to increase and, in the latter cases, the α levels are observed to decrease.

The contrast in the behavior observed with use of the AVG and the SMLE on data sets that deviate from lognormal is striking. As indicated in Figures 2A and 2B, changes in the α levels observed for decision rules employing the SMLE mimic those observed for the L-UCL$_{95}$. Thus, censoring data (Figure 2A) causes the α levels associated with use of the SMLE to decrease relative to what is observed for data that are lognormally distributed (Figure 1) and the magnitude of the effect increases both with increasing spread in the data and increasing sample size. Yet again, this suggests that statistical control deteriorates with increasing sample size. Applying the SMLE to data sets derived from P-lognormal distributions causes the α levels to increase over what is observed in Figure 1. Clearly, the SMLE statistic does *not* provide a reliable estimate of the mean of a distribution that is not lognormally distributed.

In contrast, the overall behavior of decision rules employing the AVG statistic appears to remain relatively constant over all the changes between Figures 1, 2A and 2B. The α levels obtained using the AVG for corresponding data sets (in terms of number of samples and spread in the data) across these three figures are comparable. But this simply confirms that the common arithmetic average is an unbiased estimator of the mean of *any* distribution (Box et al. 1978).

Although not depicted in the figures presented, the behavior of the T-UCL$_{95}$ was also evaluated in this study and trends in the behavior of the T-UCL$_{95}$ closely parallel those observed for the AVG. Briefly, even when applied to sample sets derived from ideal lognormal distributions, the intended statistical control is absent. This simply confirms, as reported in our first paper, that statistical procedures designed for normally distributed data cannot be expected to perform as anticipated when they are applied to data that are not adequately described by normal distributions.

The α levels observed for the T-UCL$_{95}$ are all smaller than the expected 95% and the values decrease slightly with increasing spread in the data. However, the values do increase with increasing sample size, indicating that statistical control improves with increasing sample size, as it should.

In parallel with the results observed for the AVG statistic, the pattern of α levels observed when the T-UCL$_{95}$ is applied to data sets derived from P-lognormal distributions and to censored data sets is almost identical to the pattern observed among data sets derived from lognormal distributions. Therefore, although formal statistical control is clearly lost when this summary statistic is applied to positively skewed data, its behavior appears to be insensitive to variations in the characteristics of such distributions over the range of changes that were examined in this study.

Decision Rules Incorporating Lower Confidence Limits for the Mean

The behavior of decision rules incorporating each of three statistical procedures for estimating the lower confidence limit to the mean of a distribution are compared in Figures 3A and 3B. The individual bars in Figures 3A and 3B represent the α levels for specific decision rules applied to data sets exhibiting the sample sizes and spreads indicated on the X-axis. The same underlying distributions employed for Figures 2A and 2B were employed to generate Figures 3A and 3B, respectively. What distinguishes Figures 3A and 3B from Figures 2A and 2B is the set of decision rules evaluated.

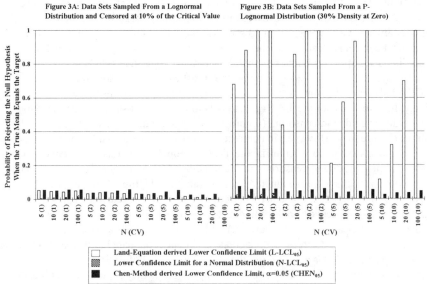

Figure 3: Alpha Values for the Indicated Decision Rules Applied to Data Sets Sampled From the Indicated Distributions

In Figures 3A and 3B, the white bars represent rules employing the $L\text{-}LCL_{05}$, the cross-hatched bars represent rules employing $T\text{-}LCL_{05}$, and the black bars represent rules employing the $Chen_{05}$ statistic. What makes the Chen method particularly interesting is that it is a new technique for deriving lower bounds for the mean of a positively skewed distribution that is designed to be insensitive to the specific shape of the high-end tail.

Effects observed on the $L\text{-}LCL_{05}$ due to censoring and due to distorting the underlying lognormal distribution by adding density at zero are apparent in Figures 3A and 3B. When applied to ideal lognormal distributions, the $L\text{-}LCL_{05}$ generally exhibits the expected statistical control so that the α levels for data sets exhibiting a range of sample sizes (5, 10, 20, and 100 samples) and a range of spreads (CV's ranging between 1 and 10) remain approximately stable at 5%. As discussed previously, however, the exact α levels appear on the whole to be slightly low.

When the data evaluated using the L-LCL$_{05}$ deviate from lognormal, the effects on the α levels for this statistic parallel those observed for the L-UCL$_{95}$ statistic. Thus, censoring the data results in a reduction in the α levels below the 5% that is expected when statistical control is maintained and the effect is largest for data sets exhibiting large spreads (with CV's of five and larger). As indicated above, however, this is simply due to the way in which censoring is performed in this study where the fraction of censored points increases dramatically with increasing spread. Similarly, evaluating data derived from P-lognormal distributions results in the α levels increasing well beyond the expected 5%.

Both effects get worse with increasing sample size (so that statistical control deteriorates with increasing sample size) and the effects are substantial. In fact, when the L-LCL$_{05}$ is applied to a P-lognormal distribution (with 30% density at zero) and a data set exhibiting a CV of 1 is examined, α levels as large as 70% are observed for data sets containing as few as five samples. With 10 samples, the α levels increase to almost 90%. Thus, statistical control is clearly lost when either Land-Equation derived upper or lower confidence limits are applied to data sets that deviate from lognormal behavior.

In the previous paper, we showed that the statistical control associated with use of the T-LCL$_{05}$ fails when this statistic is applied to lognormal distributions. Not surprisingly, therefore, use of the T-LCL$_{05}$ also fails to provide statistical control for data sets that are distorted due to censoring (Figure 3A) or that are derived from P-lognormal distributions (Figure 3B). However, in contrast to the behavior exhibited by the L-LCL$_{05}$ statistic, the α levels exhibited when using the T-LCL$_{05}$ appear to be relatively stable and independent of changes due to censoring or due to adding 30% density at zero. This can be seen by comparing the pattern exhibited for this statistic in Figure 3A and Figure 3B.

Although not depicted, the general pattern of α levels observed when applying the T-LCL$_{05}$ statistic to data sets derived from ideal (undistorted) lognormal distributions is also similar to that indicated in both Figures 3A and 3B. The α levels associated with the T-LCL$_{05}$ statistic appear to decrease with increasing spread in the data but increase with increasing sample size. At least, therefore, the statistical control of decision rules employing this statistic increase with increasing sample size, as they should. Given the relative similarity between the α levels observed for this statistic over data sets (of sizes and spreads evaluated in this study) that are derived from lognormal distributions, censored lognormal distributions, P-lognormal distributions, and censored P-lognormal distributions, the power of this test may be relatively stable over the range of characteristics likely exhibited by most environmental data sets.

Decision rules employing the Chen statistic exhibit reasonable statistical control over the data sets derived from the positively skewed distributions examined in this study (including ideal lognormal distributions, censored lognormal distributions, P-lognormal distributions, and censored P-lognormal distributions). In keeping with the designed tolerance of this test statistic, the observed α levels (over data sets of varying spread and sample number) are virtually identical no matter which of the underlying distributions are sampled and evaluated.

Importantly, however, statistical control using the Chen statistic is not perfect. Observed α levels decrease with increasing spread in the data and increase with increasing sample size. Thus, for example, the α levels observed for data sets exhibiting a CV of 1 are all close to the expected 5% for the $Chen_{05}$ statistic (i.e. the Chen statistic evaluated at the 0.05 level of significance). The α levels observed when the $Chen_{05}$ statistic is applied to data sets exhibiting a CV of 10 are approximately half the expected value (approximately 2.5%), but increase with increasing sample size. Although not shown, corresponding patterns are also observed for the $Chen_{10}$ and $Chen_{20}$ statistics (i.e. the Chen statistic evaluated at the 0.1 and 0.2 levels of significance, respectively). Importantly, because the α levels observed in association with the family of Chen statistics increase with sample size, it is expected that the power of the test also increases with sample size, as is appropriate.

Decision Rules Incorporating the Max, and the %90 and %95

Although the Max, the %90 and the %95 are not formally estimates of the mean of a data set (and are not expected to converge to the mean for arbitrarily large sample sizes), they have been used as screening tools in decision making at hazardous waste sites. Therefore, their performance is examined here. Briefly, as already demonstrated and explained in the first paper, the Max does not control Type I error very well (leading to falsely requiring cleanup with high frequency). Worse, the associated error rate increases with increasing sample size. Therefore, use of the Max test is not recommended.

As with the Max test, %90 and %95 tests are expected to represent conservative tests that are health protective. This is generally true as long as the data from which they are derived are not too skewed. Because the mean of a distribution shifts to a higher and higher percentile of the distribution as data become increasingly positively skewed, when the spread of the data become too large, percentile tests should no longer be considered conservative in a health protective sense. As indicated in the first paper, however, over the range of spreads most commonly encountered among environmental data, percentile tests may generally be considered to be protective.

Importantly, percentile tests exhibit decreases in both Type I and Type II error rates as sample sizes increase so that the power of these tests behave appropriately. We thus recommend using %90 or %95 tests for screening rather than the Max test for most applications under most conditions. Notably, both the %90 and the %95 test degenerate into a max test when sample sizes are small. Thus, one needs to analyze data containing at least 10 measurements to distinguish results for the %90 test from those for the Max test. Correspondingly, 20 measurements are required to distinguish the %95 test from the Max test. This is not a problem, however, because it is only with the larger data sets that the results of the Max test become unsupportable.

Conclusions

The results of this study indicate that none of the spatially-independent statistical procedures that have been applied to environmental data provide the advertised statistical control over the complete range of characteristics likely to be encountered among field data from hazardous waste sites. Among these tests, however, those derived for normally distributed data (i.e. the $T\text{-}UCL_{95}$ and $T\text{-}LCL_{05}$ and AVG statistics) or modified from those derived for normally distributed data (i.e. the Chen statistic for all of the α levels evaluated) do appear to provide reasonable control of error rates that, appropriately, improve with increasing sample size. Tests derived for lognormally distributed data are clearly too sensitive to deviations from lognormality to be useful generally for environmental studies.

Importantly, because the advertised statistical control is not realized, it does not appear necessary to prefer tests for which such formal control is inherent, because the required validity criteria are unlikely to be satisfied. The best approach to controlling error may therefore be to define the characteristics of the field data to be evaluated to the extent possible (given the common limitations with such data) and to select an appropriate statistical tool that has been shown to provide the best control of both types of error, for the sample size and the data characteristics that appear to bracket those of interest.

Candidate tests to be considered under such circumstances may include tests that, while not designed to provide formal statistical control, are nonetheless reasonably robust to the range of conditions that prevail. The test statistics listed in the first paragraph above appear to be about equally robust over the range of characteristics likely to be encountered among field data so that any of these might be preferred, given other project-specific requirements for the manner in which decision rules need to be defined.

Power curves that depict the relative performance of the T-UCL$_{95}$, the AVG, the Chen$_{20}$, and the Chen$_{05}$ are presented for two widely disparate sets of conditions in Figures 4A and 4B. Figure 4A is for data from a P-lognormal distribution (30% density at zero)

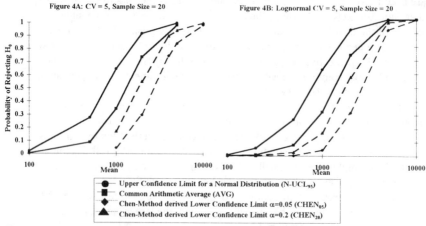

Figure 4: Power Curves for the Indicated Decision Rules Applied to Data Sampled from either a Lognormal or a P-Lognormal Distribution (30% Density at Zero) and Censored at 10% of the Critical Value

and censored at 10% of Q. Figure 4B is for a lognormal distribution with uncensored data. Both figures are for distributions exhibiting a CV of 5 with data sets containing 20 samples. These curves can be interpreted as described in the first paper.

As is apparent in both of these figures, by shifting the X-axis, the relative performance of all of these tests can be considered to be comparable (i.e. the spread in m required to go from a rejection rate of 0.2 to a rejection rate of 0.95 is between a factor of 5 and 10 in all cases (note that the X-axis is a log scale). Thus, given that the AVG is known to be an unbiased estimator of the mean for any distribution, it may potentially be the most robust of the statistics presented. The T-UCL$_{95}$ might then be considered to be a slightly more conservative (i.e. health protective) test. To achieve the right balance of Type I and Type II errors, Q needs to be defined more creatively for the various Chen statistics. Finally, given a goal of keeping Type I error at 20% when m = 0.5Q and Type II error at 5% when m = 2Q (USEPA 1996), it appears that larger sample sizes may be required for most environmental sites than those assumed in the most recent guidance. In these Figures, data sets containing 20 (uncomposited) samples are not adequate to allow the two error rate targets to be simultaneously achieved for any of the tests depicted.

Interestingly, while the results of this study largely parallel those recently reported by the U.S. EPA (1996), important differences are noted. The most striking is that the loss of formal statistical control attributed to the Chen test that was observed in this study was not reported in the U.S. EPA (1996) study. Likely, this difference is due to the differences in methodologies for the simulations performed. In the EPA simulations, samples were assumed composited prior to calculation of the various test statistics and a smaller range of CV's were examined. There also appears to be differences in the manner that censoring was handled between the two studies, although this is not clear from the writeup available at the time of our study. The cause of these differences may also be attributable to the differences in the underlying distributions examined in the two studies. Because there are times when compositing is ill advised (such as when handling contamination with volatile organic compounds or when there are other objectives requiring better information on the true variability of the field contamination), there may be cases when the results of this study may be more relevant. We also attempted to look over a larger range of sample sizes because we hoped to address broader applications than screening alone.

One final note, to minimize decision errors, it is important to assure that the manner in which data are collected in a field study match the objectives of the study. It is particularly important to pay attention to this consideration because field studies are typically designed to address multiple objectives and compromises are required. This is a crucial prerequisite to selection of the most appropriate decision rule that may determine whether there is even a chance of "getting the right answer" (Berman 1995).

References

Attfield, M.D. and Hewett, P., 1992, "Exact Expressions for the Bias and Variance of Estimators of the Mean of a Lognormal Distribution," *Amer Ind Hyg Assoc J* 53:432-435.

Berman, D.W., 1995, "Does Risk Assessment Work? Limitations Imposed on risk Assessment by Data Quality and Common Practice," *Challenges and Innovations in the Management of Hazardous Waste, Proceedings of an Air &Waste Management Association and Waste Policy Institute Conference, Pittsburgh, Pennsylvania*, pp. 493-503.

Berman, D.W., Allen, B.C. and Van Landingham, C.B., 1998, "Risk Assessment and Risk Management Under Superfund. Part 1: The Performance of Tests Used to Evaluate Environmental Data To Decide When to Remediate" *Third Symposium on Superfund Risk Assessment in Soil Contamination Studies, ASTM STP 1338* K. B. Hoddinott Ed., American Society for Testing and Materials.

Bickel, P.J. and Doksum, K.A., 1977, *Mathematical Statistics: Basic Ideas and Selected Topics*, Holden-Day, Inc., San Francisco.

Box, G.E.P., Hunter, W.G., and Hunter, J.S. 1978, *Statistics for Experimenters: An Introduction to Design, Data analysis, and Model Building,* John Wiley and Sons, New York.

Chen, L, 1995, "Testing the mean of Skewed Distributions," *J. Amer Stat Assoc,* 90:767-772.

Gilbert, R.O. 1987, *Statistical Methods for Environmental Pollution Monitoring,* Van Nostrand Reinhold, New York.

Land C.E., 1971, "Confidence Intervals for Linear Functions of the Normal Mean and Variance," *Annals of Mathematical Statistics* 42:1187-1205.

The U.S. Environmental Protection Agency, May 1992, *Supplemental Guidance to RAGS: Calculating the Concentration Term,* Office of Solid Waste and Emergency Response, Publication 9285.7-081.

The U.S. Environmental Protection Agency, 1994, *Draft Soil Screening Guidance,* Vol. 59, No. 250 (59FR67706).

The U.S. Environmental Protection Agency, May 1996, *Soil Screening Guidance: Technical Background Document* Office of Emergency and Remedial Response, OSWER-9355.4-17A.

Todd A. Wang[1], William F. McTernan[2],and Keith D. Willett[3]

A RISK BASED DECISION MODEL TO OPTIMIZE REMEDIATION CHOICES AT A SUPERFUND SITE

REFERENCE: Wang, T.A., McTernan, W.F., and Willett, K.D., **"A Risk Based Decision Model to Optimize Remediation Choices at a Superfund Site,"** *Superfund Risk Assessment in Soil Contamination Studies: Third Volume, ASTM STP 1338,* K.B. Hoddinott, Ed., American Society for Testing and Materials, 1998.

ABSTRACT: A study was initiated which combined elements of stochastic hydrology, risk assessment, simulation modeling, cost analysis and decision making to define the optimum remediation choice(s) for a Superfund site in east Texas. The underlying premise of this effort was that environmental decision making is inherently complex due to uncertainties in contaminant concentrations and resultant exposures. The technical analyst should supply the decision maker with estimates of these uncertainties as well as the cost penalties required to reduce them to manageable levels.

This study employed Monte Carlo transport modeling to define the probability of contaminant excursions from the site, applied geostatistical simulation to existing data sets, used Bayesian modeling to define the worth of additional data and Decision Modeling to define optimum configurations. These individual components were combined to produce a decision model which defined remediation alternatives given levels of risk tolerance which could be supplied by the decision maker or affected community.

KEYWORDS: risk assessment, stochastic hydrology, decision modeling, uncertainty analysis

---

Introduction

The Office of Management and Budget (OMB) estimates that approximately 61,155 toxic sites, managed by five federal agencies, may require up to 75 years for remediation. The resources required for these efforts are estimated to be between $234 and $389 billion. With various budgetary and personnel constraints, the need for optimizing resources and making sound management decisions in formulating

[1] Captain, U.S. Army Corps of Engineers, Omaha District, Rocky Mountain Area, 2032 N. Academy Blvd., Colorado Springs, Colorado 80909-1506.
[2] Professor, School of Civil and Environmental Engineering, Oklahoma State University, 207 Engineering South, Stillwater, Oklahoma. 74078-5033.
[3] Professor, Economics Department, Oklahoma State University, 339 Business, Stillwater, Oklahoma. 74078-4011.

remediation strategies is and increasingly will become more important. Management approaches to optimize resources while reducing unacceptable risk to human health and the environment must be developed for complex remediation problems given the uncertainty associated with such problems. Management issues concerning levels of risk, criteria for site evaluation, remediation alternatives, uncertainty of data and other parameters are complex. Environmental decision makers are required to incorporate these issues into a sound decision making process.

Utilizing a Decision Analysis Methodology (DAM) allows a decision maker to compartmentalize a problem and think the process through in a rational and structured manner. The DAM presented herein uses the fundamentals of decision analysis and links a decision model to stochastic modeling that quantifies economic and environmental risk.

Methodology

The purpose of the decision model is to provide the decision maker with a way of concisely visualizing the problem and a means to evaluate alternative choices within the model. A typical problem type involves choosing the best remediation alternative to minimize both cost and environmental risk.

A decision tree, as shown in Figure 1, is a convenient way to visualize the alternatives. Decision nodes (square blocks) are sequential and relate to time. Other nodes within the tree represent random events (circles) and end consequences (triangles). Branches coming from decision nodes represent courses of action. Branches following random event nodes are representative of possible states of nature. Those branches immediately prior to terminal nodes represent states of nature (consequences) for that branch of the decision tree (Ossenbruggen, 1984). Every decision within the tree leads to a consequence that has some probability of occurrence. The probability of occurrence is known occasionally but usually is uncertain and therefore estimated using stochastic modeling techniques.

Stochastic modeling can be used to develop probability distributions to address problem uncertainty. For example, transport modeling can be used to determine the probability that a contaminant will reach a specified point. The probability distributions can then be linked to a decision model that optimizes cost and risk used in the decision tree format as shown in Figure 2. That is, the transport modeling probabilities are utilized in the decision tree analysis as probabilities for the consequences.

Recent work by Freeze, Gorelick, Massmann and others (Freeze et al., 1990; Gorelick et al., 1984; Massmann et al., 1991) has been instrumental in linking uncertainty analysis to decision making and optimization. Freeze and coworkers (1990), introduced a decision analysis framework that linked hydrogeologic technical and uncertainty analysis to an economic framework that could be used in decision analysis. Critical elements of this framework were integrated into this Decision Analysis Model (DAM).

Figure 1. Decision tree used to select alternatives under uncertain conditions.

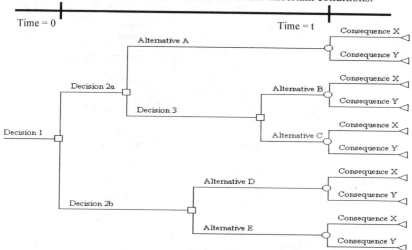

Developing the decision tree framework starts with visualizing the problem and developing a logical flow diagram of the decisions that must be made. This process is dynamic and usually requires changes or adjustments to the tree. Using a program such as Decision Analysis by TreeAge (DATA) was helpful with its ability to edit large trees quickly (DATA, 1994). Figure 3 is the decision tree developed after visualizing the example problem. The environmental decision maker is confronted with multiple decisions: take or postpone immediate action, conduct additional testing, remediate or take no action, and finally select a remedial action. Remedial actions can include passive restoration or other minimally invasive techniques.

The uncertainty of the potential for environmental risk is depicted by the chance nodes within the decision tree. Branching from the nodes are three states of nature (see Figure 2); Receptor Not Exposed to Contaminant, Receptor Exposed to Contaminant < MCL and Receptor Exposed to Contaminant ≥ MCL. Probabilities were determined for each state of nature by the appropriate analysis. For groundwater contaminants, a transport analysis using an applicable groundwater model with probabilistic input would be used to determine these probabilities. These probabilities are based upon the outputs from the stochastic contaminant transport models. These models utilizes a deterministic, public domain code linked to a Monte Carlo preprocessor. In this manner the uncertainties investigated result from "parameter uncertainty' where inherent variations in these input values are addressed by the development of cumulative density functions generally describing a central tendency as well as a variance term. The cdf's are repeatedly and randomly sampled and introduced into the deterministic model to

Figure 2. Integrating chance nodes and states of nature within the decision tree.

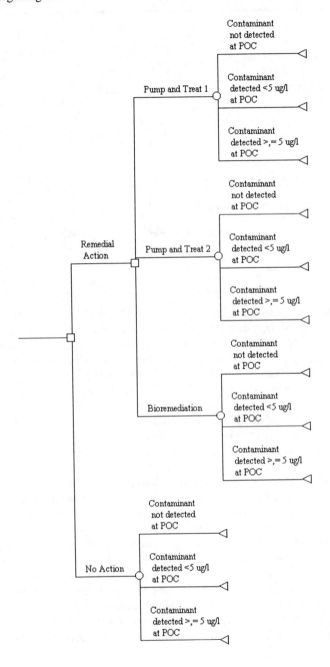

subsequently produce a series of individual computer simulations which are pooled into distinct statistical distributions which are analyzed for probabilities of occurrence. This type of uncertainty analysis does not address errors arising from an improperly configured model (model uncertainty) or from variations in the initial or boundary conditions (input uncertainties). Identification of these types of uncertainties, while important, are not routinely addressed in studies of this type.

Figure 3. Decision tree in skeletal form, depicting decision nodes within the tree.

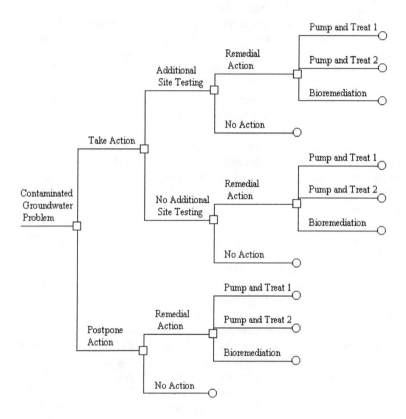

Example Problem

To illustrate an environmental decision modeling approach, a well documented contamination problem was addressed which involved an unsaturated soil loaded with a toxic compound that leached into an underlying aquifer. The site, a U.S. Army-owned munitions plant known as Longhorn Army Ammunition Plant (LHAAP) is located in Harrison County of northeastern Texas. The U.S. Army Corps of Engineer's Tulsa

District was charged with managing the project, in which a remediation phase was initiated in 1995 after years of site characterization (Murry, 1995). Corps personnel have been responsible for extensive, post-closure environmental monitoring. Wells were installed and sampled at numerous locations. The decision model developed addressed the need to develop, continue and extend a monitoring network of this type. Decisions related to the construction and operation of this network, however, were independent of this modeling exercise.

The complex, which had been secured from trespass by unauthorized individuals, is contractor-operated to load, assemble and pack pyrotechnics, illumination/signal ammunition and solid rocket propellant motors (Green et al., 1990). Since the early 1950's this industrial complex has disposed of solid and liquid explosives, pyrotechnics and combustible solvent wastes by open burning, incineration, evaporation and burial. A large portion of the wastes were disposed of in an area called "Burning Ground 3" and the now closed unlined evaporation pond (UEP) (USACE, 1993). Physical characteristics of the test contaminant compound, TCE, include high volatility, specific gravity greater than 1, moderate soil sorption properties and moderate solubility in water.

Given the situation, important concerns for the decision maker include: how much will it cost to clean up?; is there a risk to human health and/or the environment?; should there be more testing?; how much time is available to implement a plan? Figure 3 compresses these time related and dependent decisions into a concise graphic format. In this example, the first decision node defines the decision whether to *Take* or *Postpone Action*. A subsequent decision following the *Take Action* branch is the decision to perform *Additional Site Testing*. This is a logical progression as a person desires as much data as possible upon which to base a technically sound decision while lowering risk or uncertainty. By integrating the decision for additional testing into the tree, the decision maker can evaluate the worth of obtaining additional data. For example, additional testing potentially can provide additional information, which may reduce uncertainty. Is this additional testing, however, cost effective or necessary to make subsequent decisions? DAM is capable of evaluating the benefit of additional testing and providing the decision maker the optimum alternative through a Bayesian analysis.

Another subsequent decision, within the decision tree, is whether to proceed with *Remediation*? This decision is in both branches stemming from the first decision node, as shown in Figure 3. The no action alternative is a viable alternative if the risk of a contaminant reaching an environmentally significant receptor is near zero. Stochastic modeling, using Monte Carlo techniques, can quantify the uncertainty by developing probabilities that a contaminant can, for example, enter a potable water supply.

The decision tree framework is based upon the critical issues developed by the decision maker when defining the problem. Next, the consequence probabilities and costs are determined utilizing the available information and modeling tools. DAM is dynamic, thus changes to the framework as well as cost variables or probabilities are possible. This is convenient, as new information becomes available and/or priorities change the decision tree changes as well. It should be noted that the modeling output from such an exercise is an approximation based upon known data and the experience of the individuals using the model.

Formulation of a Decision Objective Function

The objective function, equation (1), was used to calculate the expected monetary value (EMV) of each decision shown in Figure 3. This objective function minimizes the cost of actions while also minimizing human and environmental exposure (risk is probability of failure). That is, the decision with the smallest EMV is the optimal decision. The EMV is a relation of objective function cost and probability for the states of nature (Ossenbruggen, 1984). For the example failure is defined as the contaminant concentration reaching a domestic groundwater well at concentrations equal to or greater than the maximum contaminant level (MCL), as given in federal and state regulations.

The first part of the objective function is a typical present-value model that was used to evaluate each alternative. The remedial cost function (C(t)) and the failure cost function (R(t)) are the core variables for this equation. Costs of additional testing and groundwater monitoring were also included. The probability function is the quantification of the problem's uncertainty, which represents the probability a certain state of nature will exist.

$$\text{Minimize } \$ = [[(1/(1+I)^t) * (C(t) + R(t) + T(t))] + M(t)] * P(sn_x) \qquad (1)$$

where
$\$$ = cost in dollars
I = index rate for money
t = time unit from 0 to T
$C(t)$ = remedial action cost function
$R(t)$ = risk failure cost function
$T(t)$ = additional testing cost function
$M(t)$ = monitoring well cost function
$P(sn_x)$ = probability for a given state of nature

Estimations of these costs were completed by application of CORA (Cost of Remedial Action), an expert system developed for the U.S. Environmental Protection Agency. U.S. EPA, 1990). Cost estimates produced by CORA were used directly in the decision model. No effort to determine uncertainties in these estimates was attempted.

Remedial Action Function

The remedial action cost function is the estimated capital and operational costs associated with each of the consequences involving remediation actions. This cost variable is a function of time; therefore, a uniform series model is used to calculate a present value for operational costs as shown below:

$$C(t) = RACAP + RAOM[(1+I)^t - 1]/[I(1+I)^t] \qquad (2)$$

where RACAP = capital cost for remedial action
RAOM = Operation and Maintenance cost for remedial action

Uncertainties in this analysis included the source characteristics and the size, location and rate of travel of the contaminant plume. Since plume characteristics are used to identify various alternative remedial actions and their attendant costs for the subsequent decision model, these estimates inherit the uncertainties associated with defining the plume. A geostatistical approach called conditional simulation was used to determine the size of the plume and its concentration distribution. A potentially powerful tool for the analysis of contaminant concentration data in groundwater, conditional simulation extends estimation methods such as kriging to produce plume simulations with attendant probabilities of occurrence. Geostatistical methods are useful for site assessment and monitoring situations where data are collected on a spatial network of sampling locations, and are particularly suited to cases where contour maps of pollutant concentration or other variables are desired (Englund and Sparks, 1991 and Cooper and Istok, 1988).

Geostatistical Analysis Method

This study utilizes publicly available geostatistical software to define the spatial statistics of the available monitoring data in a five step process. Data collection and assessment is the first step that insures quality data are used. This is followed by the variance analysis step to determine the semivariance between all the sampling points. Because many of the sampling points were spatially clustered, the third step (Sample Data Declustering) was utilized. If sample declustering weights were not assigned the collected data may be skewed and not a true representation of the entire area encompassing the sample points (Deutsch and Journel, 1992). The fifth step utilizes the data and their spatial statistics in kriging and conditional simulation algorithms using Monte Carlo techniques. This step estimates the concentration of the contaminant at unsampled locations. A simulation is "conditioned" when it honors the observed values of the regionalized variable (Hohn, 1988). Thus, a conditional simulation can be defined as a surface which has the same mean, and correlation structure variability as the studied phenomenon and passes through the observed points maintaining their values (Delhomme, 1978). This analysis results in an isoconcentration map for the contaminant plume and a measure of statistical variance around each point that is used within the decision model. A statistical analysis follows to determine the plume (two- or three-dimensional) size, location and concentration distribution within varying confidence intervals. For this analysis, the 95% confidence interval plume is defined. That is, there is only a 5% chance that the plume size and concentrations , simulated in this effort, would be exceeded. Requiring this level of confidence results in a larger plume with resultant higher remediation costs than using a simulated plume configuration with a lower confidence level. A 95%, or similar, CI is consistent with established Risk Assessment protocols and was, therefore, applied to this study.

Failure Function

The failure cost function places a monetary value on failure; therefore, it is concerned with minimizing costs. Failure can result in substantial costs due to litigation, new domestic water source development or other politically correct remedial action alternatives. For this effort only those failures resulting from runoff diversion and attendant soil excavation, site security and municipal water replacement were considered. This variable is a function of time; subsequently, the present cost of a uniform series is applied to the recurring costs associated with failure.

$$R(t) = FCAP + FOM[(1+I)^t - 1]/[I(1+I)^t] \tag{3}$$

where FCAP $=$ capital cost of failure

FOM $=$ operation and maintenance cost of failure

Additional Testing Function

The decision branch in Figure 3, ***Additional Site Testing***, utilizes Bayes theorem to provide a refined probability of occurrence for a given state of nature. The Bayesian application, in this example, utilizes data obtained from additional subsurface monitoring and testing to determine a revised probability of the contaminant migrating to a well. Capital and operational costs are incurred for additional testing. A uniform series model was used to calculate a present value for T(t), operational costs as shown below, since this variable is a time dependent function.

$$T(t) = TCAP + TOM[(1+I)^t - 1]/[I(1+I)^t] \tag{4}$$

where TCAP $=$ capital cost for testing

TOM $=$ operation and maintenance cost for testing

Monitoring Well Function

The monitoring well cost function is the operational cost for monitoring existing wells and associated laboratory analysis. This function has no capital costs because monitoring wells exist in the example evaluated. The cost variable is a function of time and a uniform series model is used to calculate a present value for operational costs as shown below.

$$M(t) = MOM[(1+I)^t - 1]/[I(1+I)^t] \tag{5}$$

where MOM $=$ operation and maintenance cost for monitoring groundwater in existing wells

Risk Function

A majority of the example's uncertainty analysis is associated with the probability or risk function [$P(sn_x)$]. For the given example, three states of nature (sn_x) are used based on the concentration of the contaminant within the groundwater reaching a receptor/domestic well. States of nature are defined as: (sn_1) no detectable amount of contaminant reaching the well, (sn_2) less than the MCL of the contaminant reaching the well and (sn_3) greater than or equal to the MCL of the contaminant reaching the well. Given these three possible events without any additional information, the probability of any one being true state of nature is 33.3%. In reality, the three states of nature most likely do not have the same probability of occurrence. Therefore, an uncertainty analysis is conducted to better define these probabilities.

The uncertainty analysis employs Monte Carlo simulation techniques to determine the probability of the contaminant reaching a receptor at a specified concentration. The probability is taken from a probability distribution function developed from this analysis. These results are used in the risk function of the decision model's objective function, equation (1), to calculate the expected monetary value (EMV) within the decision tree analysis.

The probability distributions for the contaminant concentrations at the receptor are shown in Figure 4. Three hundred Monte Carlo simulations using contaminant transport models were statistically analyzed to develop these probability distributions for chemical concentrations expected at five separate time periods within the 50 year simulation period. The analysis did not use more than 300 simulations, as it was found that maximum precision of the Monte Carlo analysis occurred between 275 and 300 simulations. This was determined by plotting mean concentrations against the number of simulations.

The resultant probabilities were developed by plotting the simulated concentrations at the receptor. Figure 4 presents the five curves illustrating the probabilities of contamination at the receptor for a specified time period. Curve A represents the probability distribution of the simulated concentrations at the receptor over the entire 50 year simulation period. Curve B shows the concentrations probabilities at the receptor for simulation years 1-10 while curve C addresses simulation years 11-20. Curves B and C show a significant difference in expected probabilities. The probability for .001 mg/l (1 ug/l) of contaminant to migrate to the receptor is about 19% in the first 10 years as opposed to 2% in the following ten years, as represented by curve C. After 20 years of transport time, the probability is almost zero, as shown by curves D and E. Probability curves B and C were important when analyzing the decision option to postpone any remedial actions.

Bayesian Analysis

The Bayesian, or probability updating, used in the model was based on Bayes theorem, which is derived from the definition of the probability of an intersection and the use of the definition of conditional probability (Ossenbruggen, 1984). The objective was to improve the probabilities for the states of nature through additional testing. The prior probability was defined as the probability that the true state of nature is the event sn_i, or $P[sn_i]$. Revised probability is equal to the conditional probability where the true state of nature is the event sn_i given the outcome of a test is Z_j or $P[sn_i/Z_j]$. The definition of the probability of an intersection and the use of conditional probability definition yields:

$$P[sn_i \cap Z_j] \qquad = \qquad P[Z_j \cap sn_i] \qquad\qquad (6)$$

or

$$P[sn_i \mid Z_j] \, P[Z_j] \qquad = \qquad P[Z_j \mid sn_i] \, P[sn_i] \qquad\qquad (7)$$

Rearranging the equation, the posterior probability is obtained.

$$P[sn_i/Z_j] \qquad = \qquad P[Z_j \mid sn_i] \, P[sn_i] \, / \, P[Z_j] \qquad\qquad (8)$$

From equation (8) it can be seen that the revised probability is a function of the prior probability. The sample likelihood $P[Z_j \mid sn_i]$ is the probability that the test event Z_j occurs given the true conditional state of nature sn_i. Utilizing the Bayes theorem equation, revised probabilities were determined given the additional information/data that could be gained from newly constructed monitoring wells. Only those potential errors associated with the laboratory analyses of samples from newly constructed wells were considered in this analysis. Bayesian updating centered upon identifying the true state of nature given measurable laboratory uncertainties. It is realized that other sources of error exist and could be included in the Bayesian updating. The technique employed could be readily expanded to include these additional sources of error should data become available.

Incorporated, the data developed by Bayesian updating into the decision analysis are shown in Figure 5. It provided the analysis with a revised probability for the states of nature but also a means to evaluate the worth of additional data. Decision tree analysis took into account the cost of procuring the additional data and provided the analysis with the results that additional testing would yield in the form of revised probabilities.

Summary of Probabilities

Probabilities for the states of nature were taken from Figure 4 for use in the decision tree analysis. Table 1 summarizes the probabilities for the states of nature under the following decision branches: *Take Action*; *Additional Site Testing*, when information became available from Bayesian analysis; and *Postpone Action*. Prior to the transport analysis the probability for any one state of nature to occur was 33.3%, since the sum of the probabilities must equal 100%. After conducting the contaminant transport modeling, the probabilities were revised to 52%, 35% and 13% for sn_1, sn_2 and sn_3, respectively, for actions taken within the first ten years of simulation. However, if *Additional Site Testing* or *Postpone Action* decisions were made, the sn_x probabilities are changed as summarized in Table 1 such that if remediation action is postponed ten

Figure 4. Contaminant transport probability for concentrations reaching a receptor.

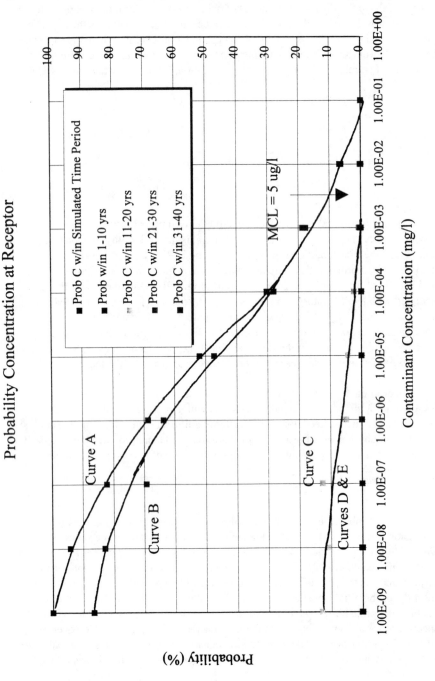

years there is only a 14% chance that the TCE concentration will exceed 5 ug/l. Additional testing lowers the probability of not detecting the contaminant to zero while producing a 33% chance that it will exceed 5 ug/l.

Figure 5. Additional Site Testing sub-branch incorporated into the overall decision tree.

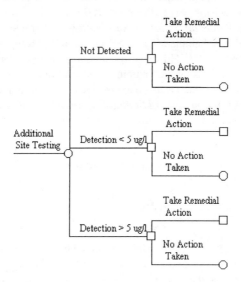

Table 1. Revised probabilities utilized within the decision tree analysis.

Decision Tree Analysis		Probability (%)		
Decision Alternative	Test Result	sn_1	sn_2	sn_3
Take Action	No additional testing	52	35	13
Additional Testing	No detection of contaminant	100	0	0
	Detection < 5 ug/l	46	42	12
	Detection ≥ 5 ug/l	0	67	33
Postpone Action 10 yrs	No detection of contaminant	100	0	0
	Detection of < 5 ug/l	95	4.5	.5
	Detection of ≥ 5 ug/l	0	86	14

Results

Figure 6 shows the results from the decision tree analysis. The optimum decisions were calculated using Expected Monetary Value theory and are highlighted in the figure by not having a double hatch on their respective branch. Given the example situation, the optimum decision was to **Postpone Action** and depending upon detection of the contaminant, either **Remedial Action** or **No Action** was warranted. If monitoring detected contamination less than the MCL, the decision of **No Action** was the optimum decision, having the lower EMV. However, if contamination detection exceeded the MCL, then **Remedial Action** using the first pump and treat remedy method had the lower EMV thus the optimum solution. Decision alternatives not optimum, as marked with a double hatch, were **Take Action** and **Additional Site Testing**. The additional testing alternative not being an optimum alternative meant that the revised probabilities (Table 1) were not cost beneficial for that decision branch to have the optimum expected monetary value. In this example problem, the Bayesian analysis revealed that additional knowledge did improve probabilities over the prior probabilities, but was not cost effective for the decision maker.

Summary

The Decision Analysis Model (DAM) presented here demonstrates its utility in environmental decision analysis under uncertain conditions. Environmental decision analysis is a very complex and dynamic process that requires a methodology that can concisely illustrate the problem and adapt to changes in conditions and uncertainty. The methodology developed provides a flexible link between uncertainty analysis and a decision optimization model. The utility of the methodology is its capability to be used in varying situations. The methodology is not limited to groundwater contamination as it was presented in the example problem. Rather, it can be applied to any environmental contamination problem that invariably involves some level of uncertainty. Had the example problem involved other types of contaminants, the decision analysis methodology would not have changed, only the tools utilized to quantify uncertainty and characterize the problem would be different.

Figure 6. Decision tree of example problem with collapsed sub-branches showing expected monetary values for decision alternatives.

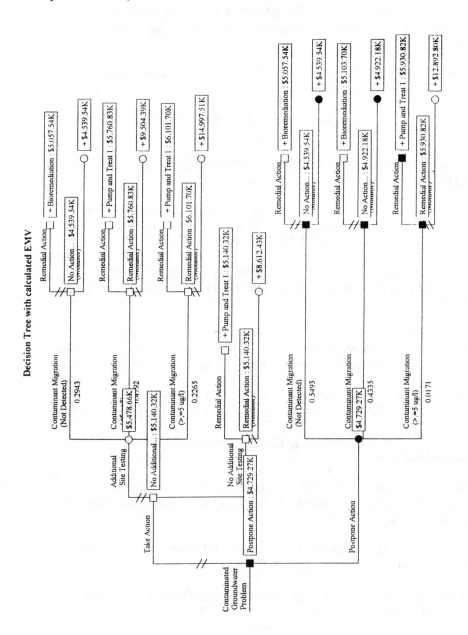

References

Cooper, R.M. and Istok, J.D. (1988). Geostatistics applied to groundwater contamination. I: Methodology. Journal of Environmental Engineering, 114(2), 270-299.

Decision Analysis by TreeAge (DATA) (1994). Decision analysis software developed by TreeAge Software, Inc.

Delhomme, J.P. (1978). Kriging in the hydrosciences. Advances in Water Resources, 1(5), 251-266.

Deutsch, C.V. and Journel, A.G. (1992). GSLIB Geostatistical Software Library and User's Guide. New York: Oxford University Press.

Englund, E. and Sparks, A. (1991). Geostatistical Environmental Assessment Software, User's Guide. Environmental Monitoring Systems Laboratory, Office of Research and Development, U.S. EPA. EPA 600/8-91/008.

Freeze, R.A., Massmann, J., Smith, L., Sperling, T., & James, B. (1990). Hydrogeological Decision Analysis: 1. A Framework. Ground Water Journal, 28(5), 738-764.

Gorelick, S.M., Voss, C.I., Gill, P.E., Murray, W., Saunders, M.A. and Wright, M.H. (1984). Aquifer reclamation design: The use of contaminant transport simulation combined with nonlinear programming. Water Resources Research. 20(4), 415-427.

Green, M.G., & Marr, A.J. (1990). Closure of an Unlined Evaporation Pond: a Case History. Bulletin of the Association of Engineering Geologists, 27(2), 235-243.

Hohn, M.E. (1988). Geostatistics and Petroleum Geology. New York: Van Nostrand Reinhold.

Massmann, J., Freeze, R.A., Smith, L., Sperling, T., & James, B. (1991). Hydrogeological Decision Analysis: 2. Applications to groundwater contamination. Ground Water Journal, 29(4), 536-548.

Murry, Cliff (1995). Personal conversation with Cliff Murry (Project Engineer for the U.S. Army of Corps of Engineers, Tulsa District).

Ossenbruggen, P.J. (1984). Systems Analysis for Civil Engineers. New York, New York: John Wiley and Sons.

References (continued)

U.S. Army Corps of Engineers (USACE) Tulsa District (1993). Data summary report of investigation results from 1976 through 1992 for Burning Ground 3 & the unlined evaporation pond. Longhorn Army Ammunition Plant, Karnack, Texas.

Nii O. Attoh-Okine[1]

POTENTIAL APPLICATION OF INFLUENCE DIAGRAM AS A RISK ASSESSMENT TOOL IN BROWNFIELDS SITES

REFERENCE: Attoh-Okine, N. O. **"Potential Application of Influence Diagrams as a Risk Assessment Tool in Brownfields Sites,"** *Superfund Risk Assessment in Soil Contamination Studies: Third Volume, ASTM STP 1338,* K. B. Hoddinott, Ed., American Society for Testing and Materials, 1998.

ABSTRACT: Brownfields are vacant, abandoned, or underutilized commercial and industrial sites and facilities where real or perceived environmental contamination is an obstacle to redevelopment. These sites are vacant because they often do not meet the strict remediation requirements of the Superfund Law. The sites are accessible locations with much of the infrastructure, albeit deteriorated, in place. Thus they also represent an opportunity to slow down suburban and rural sprawl. As a liability, the concern stems from the environment liability of both known and unknown site contamination. Influence diagrams are tools used to represent complex decision problems based on incomplete and uncertain information from a variety of sources. The influence diagrams can be used to divide all uncertainties (Brownfields site infrastructure impact assessment) into subfactors until the level has been reached at which intuitive functions are most effective. Given the importance of uncertainties and the utilities of the Brownfields infrastructure, the use of influence diagrams seem more appropriate for representing and solving risks involved in Brownfields infrastructure assessment.

KEYWORDS: Brownfields, influence diagrams, risk, infrastructure

---

Brownfields are vacant, abandoned, or underutilized commercial and industrial sites and facilities where real or perceived environmental contamination is an obstacle to redevelopment. These sites lie somewhere between significantly contaminated sites (Superfund Sites) and pristine Greenfields. Most Brownfields are concentrated in the Northeast and Midwest where much of the economy was historically based on heavy industrial activity. However, Brownfields are also common in the South and West and represent a wide variety of past industrial and commercial uses.

In 1980, the *Comprehensive Environmental Response, Compensation and Liability Act* (CERCLA), also known as Superfund, was enacted to facilitate the clean up of heavily

---

[1]Assistant Professor., Dept. Of Civil Eng. FIU   Miami, FL 33199

contaminated sites nationwide. Administered by the United States Environmental Protection
Agency (USEPA), the Superfund process was designed to establish an inventory of hazardous
waste sites nationwide, known as the National Priority List, and to transfer the cost of cleaning
up these sites to the producers of the waste.

In 1986, the *Superfund Amendments and Reauthorization Act* (SARA) established a fund,
derived from taxes on the chemical manufacturing industry, to provide for cleanup of
contaminated sites whose generators could not be determined. However, litigation and the fear
of liability kept remediation and redevelopment of the majority of these sites at a standstill.
Indeed, the Congressional Budget Office has reported that each site on the National Priority List
requires 12 years and 30 million dollars to clean up, with eight of those years and 36 percent of
that money going to the litigation process [1]. Moreover, several developers and businesses
found the federal Superfund program complicated and unable to provide risk assessments and
immunity from further environmental liability once the sites had been cleaned up. Thus, the very
Superfund process that was created to eliminate contaminated sites, seemed to result in an
indefinite preservation of the contaminated status of the majority of these sites.

As long as these contaminated sites remain unaltered, their obvious health and environmental
hazards also lingered. In addition, they negatively affected the overall economic and social
health of the communities surrounding them, a situation which made the site even less attractive
for redevelopment. These contaminated sites have therefore created environmental, economic
and social drawbacks for localities, regions, states, and hence the nation at large. Locally,
contaminated sites have contributed to blighted neighborhoods with declining central business
districts, stigmatized by decaying and abandoned sites with little potential for attracting business.
Regionally, increasing development at urban fringes has resulted in uncontained urban sprawls,
with its associated infringement on virgin Greenfields. This widespread increase in urban
sprawls have emphasized the advantage of developing Brownfields rather than Greenfields. As
Barnette [2] points out:
1.   Brownfields are properly zoned and thus well suited for industrial (and commercial) use;
2.   The civil infrastructure and utilities necessary for industrial operations are already in
        place at several Brownfield sites;
3.   Brownfield redevelopment preserves the nation's virgin land and natural resources.

Thus, while the cleanup of the site contamination seemed to fall squarely on the shoulders of the
USEPA in the early 1980s, the late 80s and the early 90s have proven rather different. Local
entities have come to realize and demonstrate that they have a significant stake in eliminating
contaminated sites from their neighborhoods.

Brownfield Redevelopment Process
While different Brownfield redevelopment programs may be tailored to suit specific needs and
objectives of different localities, most Brownfields programs seem to follow a generic process
involving the following four basic steps:
(i)   Site Identification,
(ii)   Site Assessment,
(iii)   Site Remediation, and

(iv)   Site Redevelopment.
The application of the influence diagram will only focus on site assessment.

Site Assessment

If the Phase I assessment reveals evidence of contamination, a Phase II level assessment may then be conducted. This includes actual sampling of the soil and groundwater, and results in a determination of the actual type and extent of the site contamination. This phase also involves the determination of appropriate cleanup standards, the identification of feasible site remediation technologies for cleaning up the contamination, and an estimation of site remediation costs. Determination of a feasibility plan and level of cleanup is based on a host of criteria including toxicity, exposure pathways and associated risk, surrounding land uses, economic considerations and future land use(s). The decision to proceed with site assessment will involve some levels of legal liability and financial assurances, as well as favorable socio-economic factors. In cases where a number of sites are to be redeveloped, an attempt may be made to prioritize the redevelopment of sites in an order that makes the best use of usually limited resources.

Influence Diagrams

Influence diagrams [3] provide a graph-theoretic framework for modeling probabilistic dependence of information between uncertain variables, decision options and utility functions in complex decision systems. Influence diagrams offer an important complement to more traditional representations such as decision trees and tabular listings of joint probability distribution and outcome for each action. An influence diagram is often less complex than a decision tree where branches can easily reach into the hundreds. An influence diagram is more easily constructed and more easily expanded [4]. Influence diagrams provide an explicit representation of probability dependence and independence in a manner accessible to decision makers and computers. The representation eases the assessment of coherent prior distribution. Knowledgeable engineers and experts may also express and understand more general kinds of dependence and independence assumptions.

An influence diagram is an acyclic graph whose nodes are connected with directed arcs. The different kinds of nodes have varying kinds of information stored in them. The four kinds of nodes are:
a) Probabilistic (also called Chance): an uncertain variable in a model.

b) Deterministic: a variable whose value is determined given the values of its condition variables.

c) Deterministic: a variable whose value is under the control of the decision maker.

d) Expected Value (also called Payoff): the objective function.

Probabilistic nodes are drawn as circles, deterministic nodes as double rounded circles, decision nodes as rectangles, and expected value nodes as double rectangles (Figure 1).

A directed arc into a decision node or expected value node is called an information arrow. A

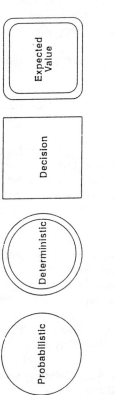

Figure 1: Nodes

directed arc into a probabilistic or deterministic node is called a relevance arrow. Further interpretation of node/arrow configurations are listed in Figure 2. Arcs leading into decision nodes represent information available at the time the decision is made. Arcs leading into probabilistic nodes indicate that the likelihood of the chance event depends (or is conditional) on the outcome at the decision or probability node from which the arc originated. "No forgetting" arcs are placed between decision nodes to signify that decisions are sequential in time, and the value of a past decision is remembered. In general, arrows in influence diagrams obey the following rules:

Rule 1: There are no directed circles in the graph.
Rule 2: The decision node are linearly ordered by information arrows.
Rule 3: Each decision node inherits any information arrows into preceding decision nodes.
Rule 4: There are no arrows from the expected value node.

The mathematical interpretation between state nodes can be deterministic or probabilistic. Assignment of probability distributions to the influence diagram depends on the structure of the nodes and relevances.

Smith [5] outlined three advantages of influence diagrams over decision trees. First and foremost, an influence diagram provides a much more compact description of a problem than does the corresponding decision tree. Second, it represents the relationship between variables equally, whether or not those variables are discrete, continuous or a mixture of both. Thirdly, it has the advantage of being able to represent succinctly conditional independence between variables in a problem.

Consider two variables A and B. In the example represented in Figure 3a, the discrete probability is a vector $\{A_i|C\}$; brackets enclose the probability assessment. "A" represents the variable to which the probability assessment will be assigned and C represents the state of information. The relevance is represented in the matrix, $\{B_j|A_i\ C\}$. The product of the unconditional vector and conditional matrix results in the unconditional vector that is assigned to $\{B_j|C\}$: $\{A_i|C\} \times \{B_j|A_i,C\} = \{B_j|C\}$. In Figure 3b, "B" conditioned on the state of information, C can be determined as follows: $\{B_j|C\} \times \{A_i|B_j,C\} = \{A_i|C\}$.

With respect to a given node, the following definitions are of importance:
a.   A *path* from one node to another is a set of arrows connected head to tail that forms a directed line from one node to another.
b.   Th *predecessor set* of a node is a set of all nodes having a path leading to the given node.
c.   The *direct predecessor set* of a node is the set of nodes having an arrow connected directly to the given node.
d.   The *indirect predecessor set* of a node is the set formed by removing its direct predecessor set from its predecessor set.
e.   The *successor set* of a node is the set of all nodes having a path leading from the given node.
f.   The *direct successor set* of a node is the set of nodes having an arrow connected directly from the given node.

- The decision maker knows the outcome of probabilistic variable X when decision Y is made

- The probability of variable Y depends on decision X.

- The decision maker knows decision X when decision Y is made.

- The probabilities associated with variable Y depend on the outcome of probabilistic variable X.

- The probability associated with probabilistic variable Y is independent of probabilistic variable X.

Figure 2: Interpretation of Arrows

g.   The *indirect successor set* of a node is the set formed by removing all elements of its direct successor set from its successor.

The nodes of the influence diagram can be numbered, say X1, X2,...Xm so that i < j whenever there is arrow (information arrow or relevance arrow) from Xi to Xj. (This means that i < j whenever Xi is a predecessor of Xj). Such a numbering is called influence numbering. Most influence diagrams have more than one numbering. The existence of an influence numbering makes clear to a probabilist why choosing a decision node determines a joint probability distribution for all the variables in an influence diagram. These decisions rules together with conditional probabilities for the other nodes (probabilistic and deterministic), amount to a specification of a conditional probability for each variable given a subset of the preceding variables in numbering. Multiplying these conditional probabilities produces a joint probability distribution. Figure 4 outlines the generic levels in influence diagram applications.

Solution
Optimization is performed by maximizing the expected value which represents the net benefit. The influence diagram has more than one solution. An assignment of a decision to each decision node is called a solution of the influence diagram if it results in a joint distribution that maximizes the expected value. In influence diagrams, a decision problem is solved by sequentially eliminating all whole nodes from the diagram, except the terminal value node, while recording the optimal policy choices as the decision nodes are removed. A randomized value for a decision variable turns into a chance variable instead of a deterministic variable. Sometimes randomized decision rules are proposed, instead of deciding what to do given each configuration of the known variables at that point. Different types of nodes are evaluated (reduced) according to the following principles. For a deterministic node, the distribution is propagated and the reduced node's distribution is substituted in its successor's distribution. Node removal and arc reversals are used in the evaluation of a probabilistic node [6]. When a node is removed, it can be dropped from the current node set, and all arcs incident on it can be dropped from the current arc set. The arc reversal transformation is Bayes Theorem. The following two theorems elaborate the points explained above.

Node Removal
Given that chance node i directly precedes the value node and nothing else in an oriented, regular diagram, node i may be removed by conditional expectation. Afterwards, the value node inherits all of the conditional predecessors from node i, and thus the process creates new barren nodes.

Arc Reversal
Given that there is an arc (i, j) between chance nodes i and j, but there are no directed (i, j)- path in a regular influence diagram, arc (i, j) can be replaced by arc (j, i). Afterwards, both nodes inherit each other's conditional predecessors.

Sensitivity Analysis and Value of Perfect Information
The decision analytic concept of expected value of perfect information is defined as the difference between expected values of a decision problem with and without the perfect information. If the expected values of information is greater than the cost of obtaining the

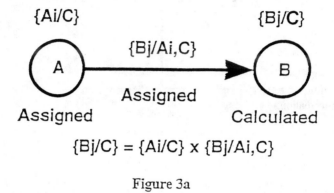

$$\{Bj/C\} = \{Ai/C\} \times \{Bj/Ai,C\}$$

Figure 3a

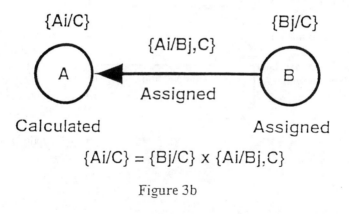

$$\{Ai/C\} = \{Bj/C\} \times \{Ai/Bj,C\}$$

Figure 3b

Figure 3: Assigned Probabilities

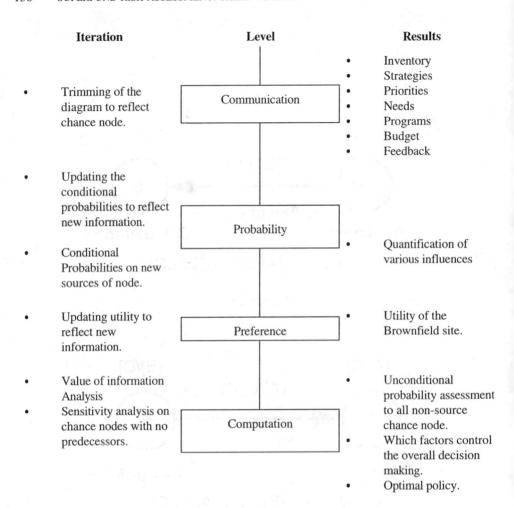

Figure 4.   Levels in Influence Diagrams of a Brownfield Site Assessment.

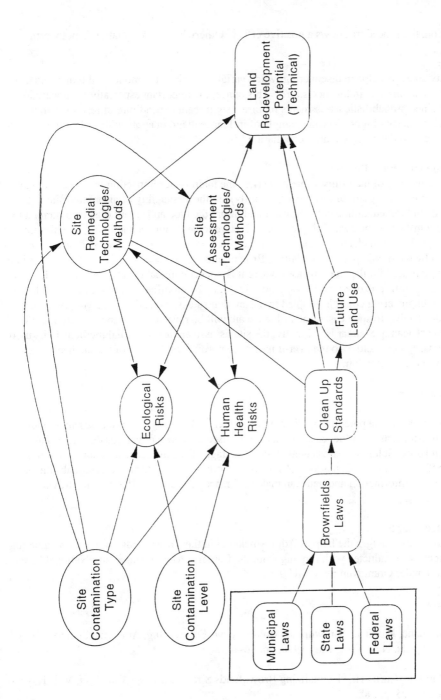

Figure 5: Technical Issues in Brownfields Remediation and Redevelopment

information, the procedure shows a positive effect. Otherwise the information is not worth obtaining.

Sensitivity analysis helps in discovering which variables "drive" the model and it answers the question "What matters in this decision?" This concept is important especially for uncertain variables since probabilistic assessments require a significant expenditure of resources to be performed properly. Figure 5 is an example of influence diagram applications of technical issues in Brownfields redevelopment and remediation [7].

Modeling of Technical Issues
The main attributes of the utility value node (i.e., the main issues impacting the expected value of the site from the viewpoint of technical feasibility) are the availability of affordable site assessment and remediation technologies. In addition, the type and extent of site contamination will largely influence the availability of cost-effective remediation technologies, and also impact on the adopted level of cleanup standards for the site. Site cleanup standards will also be influenced by the municipality's prevailing Brownfields-related laws as well as the viable future land uses permitted for the site. In cases where it is not economical to achieve feasible levels of cleanup that permit prezoned land uses, land use(s) may be modified to suit clean up levels and land use or engineering controls adopted to ensure that such sites retain the newly prescribed land use(s) indefinitely. In such cases, the influence arc will be reversed from the "Future Land Use" node to the "Cleanup Standards". Finally, the site assessment and remedial methods to be used will necessarily be sensitive to the ever-important environmental issues of human health and natural resources protection.

Summary

This paper presents the potential application of influence diagrams as a risk assessment tool. The example illustrates how influence diagrams can be used to represent various decisions and chances in Brownfields site assessment. One advantage of the influence diagram is that it can capture both qualitative information (graphical) and quantitative information (probabilities). Availability of information and data can make influence diagrams another tool in Brownfields site assessment.

Acknowledgments
The author acknowledges the help of Adjo Amekudzi in the Brownfields information gathering. This project was conducted with funding from the Carnegie Mellon University/National Science Foundation under Grant Number CMU 06975.

References

[1]   Moldano, Monica., "Brownfields Boom." *Civil Engineering*, American Society of Civil Engineers, New York, May 1996, pp. 36-40.

[2]   Barnette, Curtis H., "Revitalizing Brownfields Sires." *Iron Age New Steel*, Vol. 1, No. 6, June 1995, pp. 88.

[3]   Howard, R. A. and Matheson, J. E., "Influence Diagrams." In Howard and Matheson eds. The Principles and Application of Decision Analysis, Menlo Park, California, Vol. 2, 1984,   pp. 721-762.

[4]   Hong, Yaun and Apostolakis, G., "Conditional Influence Diagrams in Risk Management," *Risk Analysis*, Vol. 13, No. 6, 1993, pp. 625-636.

[5]   Smith, J. Q., "Influence Diagrams for Bayesian Decision Analysis," *Operational Research*, Vol. 40, 1989, pp. 363-376.

[6]   Shachter, R. D., "Evaluating Influence Diagrams," *Operations Research*, Vol. 34, No. 6, 1986, pp. 871-882.

[7]   Amekudzi, A. and Attoh-Okine, N., "Conceptual Frameworks for Understanding Brownfields Redevelopment Issues," *Infrastructure*, John Wiley & Sons, Inc., Vol. 2, No. 2, 1997, pp. 45-59.

Summary

This Special Technical Publication (STP) on Risk Assessment in Soil Contamination Studies serves to present a sampling of the state of the art and the leading edge of research in the field. This STP was not intended to be a step by step outline of the EPA's basic rules for conducting a risk assessment, rather it represents modifications to the basic method which have been acceptable to the EPA at specific sites. Therefore, this STP complies with the EPA's written desire not to inhibit research in this area and stagnate the field.

The papers presented in this STP can be grouped into the following categories:

- background determination and statistics,
- ecological risk, and
- risk management

The papers included herein have undergone peer review and extensive revision since their original presentation and provide state of the art information on conducting risk assessments for complex sites. This collection should not be viewed as the sum total of all the method modifications suitable for risk assessment use or as a guarantee of EPA acceptance, but rather as a starting point from which the environmental professional can begin to realize the wealth of variability available in assessing health risk from a site.

Background Determination and Statistics

The determination of the background condition is fundamental to establishing if an area is contaminated. At the onset of environmental risk assessment, this function was handled by one sample or referring to a national publication. In today's risk assessments, background determinations are as sophisticated as the site characterization. As funding becomes tighter, there is pressure to reduce the amount of data used to make decisions. The use of statistical methods to determine the magnitude of differences in populations of data is becoming more sophisticated. The incorporation in the symposium and in this technical publication is indicative of the importance of the methods proposed in these papers in an environmental evaluation.

Cook presented a method to use graphic statistics to minimize the sampling needs to determine background concentrations for inorganic analytes.

Ball and Hahn used a simulated lognormal distribution to test the accuracy and stability of the statistical method recommended in Superfund guidance and in RCRA SW 846.

Ecological Risk

Assessment of the non-human receptors of environmental contamination has been the deciding parameter in many site remediations. These papers attempt to provide insight to evaluating these effects whether the non-human receptor has a human connection or not.

Burns, Cornaby, Mitz, and Hadden used probabilistic methods with ecological hazard quotients to clarify the meaning of the quotient and highlight the importance of the exposure level parameter and the effects-threshold levels.

Linder focused on wetland food chains and terrestrial food chains to illustrate an approach for the derivation and validation of trophic transfer factors for metals considered chemicals of potential concern.

Risk Management

Risk assessments are basically manipulations of the data collected from the field and analyzed in a laboratory. Once the final risks have been calculated, it is the job of risk management to determine how to minimize the risk to the exposed receptors. This can take the form of removals, engineering controls, institutional controls, or in some cases no action. It is also the responsibility of risk managers to determine if action taken to reduce risk would be more environmentally damaging or result in more chemical exposure than performing no remedial action at all.

Attoh-Okine presented a methodology to use influence diagrams to manage the environmental liability of reusing Brownfield industrial sites.

Lee and Lee discuss an alternative approach to monitoring storm water runoff that shifts the monitoring program from periodic storm water sampling to analysis of the receiving waters to determine what, if any, water quality use impairments are occurring.

Mohr and Illich outline a process to expedite risk characterization at hazardous waste sites in order to establish the basis for evaluating remedial alternatives in various environmental media.

Wang, McTernan, and Willett used Monte Carlo transport modeling to define the probability of contaminant excursions from the site, employed geostatistical simulation to evaluate existing data sets, used Bayesian modeling to define the worth of additional data, and decision modeling to define the optimum configurations.

Berman, Allen, and Van Landingham compared existing statistical methods to determine if current procedures for conducting risk assessments help to make the right decisions concerning the need for cleanup.